Student Solutions Manual and Notetaking Guide

Discovering Mathematics
A Quantitative Reasoning Approach

Richard N. Aufmann
Palomar College

Prepared by

Christine S. Verity

CENGAGE

Australia • Brazil • Mexico • Singapore • United Kingdom • United States

CENGAGE

For product information and technology assistance, contact us at **Cengage Customer & Sales Support, 1-800-354-9706 or support.cengage.com**.

For permission to use material from this text or product, submit all requests online at **www.cengage.com/permissions**.

ISBN: 978-0-357-02806-3

Cengage
20 Channel Street
Boston, MA 02210
USA

Cengage is a leading provider of customized learning solutions with employees residing in nearly 40 different countries and sales in more than125 countries around the world. Find your local representative at: **www.cengage.com**.

Cengage products are represented in Canada by Nelson Education, Ltd.

To learn more about Cengage platforms and services, register or access your online learning solution, or purchase materials for your course, visit **www.cengage.com.**

Printed in the United States of America
1 2 3 4 5 26 25 24 23 22

TABLE OF CONTENTS

OPTIONAL CHAPTERS

NOTE: DEPENDING ON WHAT YOUR INSTRUCTOR HAS CHOSEN TO INCLUDE IN YOUR COURSE, YOU MAY OR MAY NOT BE USING ANY OF THE FOLLOWING CHAPTERS

NOTETAKING GUIDE

Chapter 1: Introduction to Problem Solving

THINK ABOUT IT 1.1

1. Inductive

2. Specific

3. One example is 5, which is an odd number.

EXERCISE SET 1.1

1. 64. The numbers are the squares of consecutive integers. $8^2 = 64$.

3. $\frac{13}{15}$. Add 2 to the numerator and denominator.

5. –13. Use the pattern of adding 5, then subtracting 10 to obtain the next pair of numbers.

7. Each image has a smaller square inside a larger square. The smaller square moves to a new position in a counterclockwise direction. The next figure is shown.

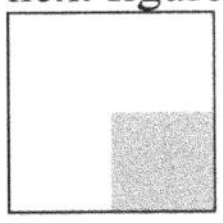

9. Each image has a smaller square and a circle inside a larger square. The smaller square moves to a new position in a corner in a counterclockwise direction. The circle moves to a new position in a counterclockwise direction. The next figure is shown.

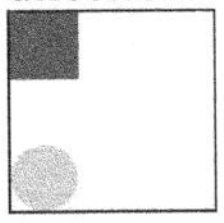

11. Each figure is a circle with a polygon, alternating positions inside and outside. The first figure is a triangle (3 sides) with a circle inside. The second is a circle with a square (4 sides) inside. The next figure is shown.

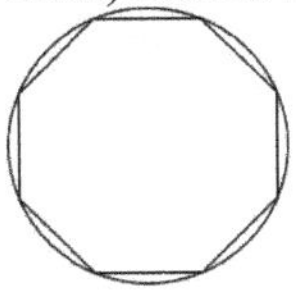

13. The amount is decreasing by $1000 per month. Thus, the amount in year 6 will be $5000.

15. More than. The increase in average annual salary increases from $7763 to $8927 to $10,266. Thus the increase from 20 years to 25 years will be more than $10,266, giving an average annual salary for a teacher having 25 years of experience of greater than $78,705 + $10,266 = $88,971, which is greater than $88,000.

17. For every 10 seconds, the distance is increasing by 300 feet. Therefore, after 70 seconds, the athlete will run a distance of 2100 feet.

19. From the table, as the depth increases, the temperature decreases. Since the last entry of the table is 40 m and 11°C, then for 45 m, the temperature should be less than 11°C.

21. From the table, the distance (in feet) is equal to the time (in seconds) squared. So at 7 seconds the distance is 49 feet, and at 8 seconds the distance is 64 feet.

23. Answers will vary. For instance, $\frac{1}{4} \times \frac{1}{2} = \frac{1}{8}$.

25. A diamond shape can have four equal sides and is not a square.

27. Whales are mammals and do not have legs.

29. a. Deductive
    b. Inductive

31. Place 4 coins in the left balance pan, 4 coins in the right balance pan, and set 4 coins aside. There are three possibilities. The right side goes down, the left side goes down, or the scale balances (the heavier coin is set aside).
Take the 4 coins that include the heavier coin and place 2 on the left balance pan and 2 on the right balance pan.
From the 2 coins on the heavier pan, place one coin on each balance pan to determine the heavier coin.

33. Label each stack of coins as 1, 2, 3, ... 10. From each stack select the same amount of coins as the labels, so select 1 coin from stack 1, 2 coins from stack 2, and so on to 10 coins from stack 10. Weigh the selected coins. Since the counterfeit coins each weigh 0.1 g more, the multiple of 0.1 g over the weight expected will determine which stack contains the counterfeit coins.

35.

	Util	Auto	Tech	Oil
A	X1	X1	✓	X3
T	X1	X1	X3	✓
M	✓	X2	X3	X3
J	X2	✓	X2	X2

Maria: the utility stock; Jose: the automotive stock; Anita: the technology stock; Tony: the oil stock

37.

	Coin	Stamp	Comic	Baseball
A	X3	✓	X3	X4
C	X2	X4	X1	✓
P	✓	X3	X3	X2
S	X2	X3	✓	X3

Atlanta: stamps; Chicago: baseball cards; Philadelphia: coins; San Diego: comic books

THINK ABOUT IT 1.2

1. a. Estimate
   b. Estimate
   c. Exact
   d. Estimate
   e. Exact
   f. Exact

2. $5.32 \times 10^{-6} = 0.00000532$
   The answer is b, greater than 0 but less than 1.

3. True

4. False; $10^{-3} = 0.001 = \dfrac{1}{1000}$

5. Yes

EXERCISE SET 1.2

1. $1487 \text{ miles} \approx 1500 \text{ miles}$
   $\dfrac{1500 \text{ miles}}{3 \text{ days}} = 500 \text{ miles per day}$

3. $2(25) + 30 = 80$
   So, $25°\text{C} \approx 80°\text{F}$.

5. $400 \times 20 = 8000$ square feet

7. Answers will vary depending on the grid. Using a 3 by 5 grid, we count 30 cars. Multiply 30 by the number of sections, 15, to get 450. So our estimate is 450 cars.

9. Answers will vary depending on the grid. Using a 3 by 5 grid, we count 18 cadets. Multiply 15 by the number of sections, 15, to get 225. So our estimate is 225 cadets.

11. $\dfrac{6,400,000 \text{ square feet}}{4 \text{ square feet}/(1 \text{ person})} = 1,600,000 \text{ people}$
    or 1.6 million people

13. If a 10-foot pole casts a shadow of 18 feet, and a telephone pole has a shadow of 72 feet, then the telephone pole will be more than half as high as its shadow is long. The telephone pole has height
    $$72 \times \frac{10}{18} = 40 \text{ feet}$$

15. $0.0000000000089 \text{ meter} = 8.9 \times 10^{-12} \text{ meter}$

17. $35,000,000,000,000 \text{ cells} = 3.5 \times 10^{13} \text{ cells}$

19. Yes. The higher the advanced degree, the higher the bar of the lifetime earnings is.

21. Yes. The Dogs category is more than half of the pie graph.

23. a. The graph crosses \$1 trillion (or \$1000 billion) in years 2013 to 2014.
    b. The graph crosses \$1.2 trillion in year 2016.

THINK ABOUT IT, SECTION 1.3

1. Understand the problem, devise a plan, carry out the plan, review the solution

2. No. We would need to know the speed at which she ran.

EXERCISE SET 1.3

1. Let g be the number of first grade girls, and let b be the number of first grade boys. Then $b + g = 364$ and $g = b + 26$. Solving gives $g = 195$, so there are 195 girls.

3. There are 36 1 × 1 squares, 25 2 × 2 squares, 16 3 × 3 squares, 9 4 × 4 squares, 4 5 × 5 squares and 1 6 × 6 square in the figure, making a total of 91 squares.

5. Solving:
   x = cost of the shirt
   $x - 30$ = cost of the tie
   $$\begin{aligned}(x-30)+x &= 50\\ 2x-30 &= 50\\ 2x &= 80\\ x &= 40\end{aligned}$$
   The shirt costs \$40.

7. There are 14 different routes to get to Fourth Avenue and Gateway Boulevard and 4 different routes to get to Second Avenue and Crest Boulevard. Adding gives that there are 18 different routes altogether.

9. Try solving a simpler problem to find a pattern. If the test had only 2 questions, there would be 4 ways. If the test had 3 questions, there would be 8 ways. Further experimentation shows that for an n question test, there are 2^n ways to answer. Letting $n = 12$, there are $2^{12} = 4096$ ways

11. 8 people shake hands with 7 other people. Multiply 8 and 7 and divide by 2 to eliminate repetitions to obtain 28 handshakes.

13. Let p be the number of pigs and let d be the number of ducks. Then $p + d = 35$ and $4p + 2d = 98$. Solving gives $d = 21$ and $p = 14$, so there are 21 ducks and 14 pigs.

15.

Dimes	Nickels	Pennies
0	0	25
0	1	20
0	2	15
0	3	10
0	4	5
0	5	0
1	0	15
1	1	10
1	2	5
1	3	0
2	0	5
2	1	0

There are 12 ways.

17. The units digits of powers of 4 form the sequence 4, 6, 4, 6,.... Even powers end in 6. Therefore, the units digit of 4^{300} is 6.

19. The units digits of powers of 3 form the sequence 3, 9, 7, 1, 3, 9, 7, 1, ... Divide 412 by 4 to obtain the remainder 0, which corresponds to 1. Therefore the units digit of 3^{412} is 1.

21. Draw a simpler picture:

Start Finish

Note that the first page of the first volume is the second dot on the line, and the last page of the third volume is the seventh dot on the line. This is because when books sit on a shelf, the first pages are on the right side of the book and their last pages are on the left side.

$\frac{1}{8}+\frac{1}{8}+1+\frac{1}{8}+\frac{1}{8}=1\frac{1}{2}$ inches.

23. If n is the number of columns of tile on the floor, then there are n blue tiles on the diagonal that starts in the upper left hand corner. Similarly, there are n blue tiles in the diagonal that starts in the upper right hand corner. The two diagonals have one tile in common, so the actual total number of blue tiles is $2n - 1$. Since $2n - 1 = 101$, we can solve to find $n = 51$. The total number of tiles is n^2. Substituting the value for n yields 2601.

25. Let b be the number of boys in the family and g be the number of girls. The first two statements imply that the speaker is a girl. Thus, $g - 1 = b + 2$. Solving for b, $b = g - 3$. To answer the last question, we must omit the youngest brother, so $b - 1 = g - 4$. There are four more sisters than brothers.

27. The bacteria population doubles every day, so on the 11th day there are half as many bacteria as on the 12th day.

29. Let x be the score that Dana needs on the fourth exam:

$$\frac{82+91+76+x}{4}=85$$
$$249+x=340$$
$$x=91$$

31. a. Place four coins on the left balance pan and the other 4 coins on the right balance pan. The pan that is the higher contains the fake coin. Take the four coins from the higher pan and use the balance scale to compare the weight of two of these coins to the weight of the other two coins. The pan that is the higher contains the fake coin. Take the two coins from the higher pan and use the balance scale to compare the weights. The pan that is the higher contains the fake coin. This procedure enables you to determine the fake coin in 3 weighings.

b. Place 3 of the coins on one of the balance pans and 3 coins on the other balance pan. If the pans balance, then the fake coin is one of the two remaining coins. You can put each one of these coins on a balance pan and the higher pan contains the fake coin. If the 3 coins on the left do not balance with the 3 coins on the right, then the higher pan contains the fake coin. Pick any 2 of these 3 coins and use the balance scale to compare their weights. If these 2 coins do not balance, then the higher pan contains the fake coin. If these two coins balance, then the 3rd coin (the one that you did not place on the balance pan) is the fake. In either case this procedure enables you to determine the fake coin in 2 weighings.

33. The correct answer is a., 1600. Sally likes perfect squares.

35. The correct answer is d., 64. The numbers are all perfect cubes. The missing number is the cube of 4.

Think About It, Section 1.4

1. 25

2. Assuming Jeremiah divides the land as required by the rules of fair division, this is a fair division of the land. The size of a piece does not matter: the value of the piece is the important fact.

3. Craig. The divider must divide an asset so that each piece has the same value.

4. $P_1 : S_1,\ S_3;\quad P_2 : S_1,\ S_3;\quad P_3 : S_1,\ S_2, S_3$

Exercise Set 1.4

1. a. $\$9 + \$6 + \$3 = \$18$

   b. One-third of the cheese, one half of the pepperoni, and none of the green pepper is $\frac{1}{3}\cdot\frac{1}{3}+\frac{1}{2}\cdot\frac{1}{3}+0\cdot\frac{1}{3}=\frac{5}{18}$, which is a lot less than half of the pizza.
   No. She does not receive a fair share.

3. Answers may vary. The easiest solution is to cut the cake so that each piece is one-half strawberry and one-half chocolate.

5. Yes. Since Amanda values the chocolate icing at \$12 and the strawberry icing also at \$12, dividing the cake into three equal pieces, regardless of icing, causes each piece to be a fair share.

7. Answers may vary. Here is one possibility.

	S_1	S_2	S_3	Bids
Caroline	45%	25%	30%	S_1
Jasmine	40%	30%	30%	S_1
Jean	$33\frac{1}{3}\%$	$33\frac{1}{3}\%$	$33\frac{1}{3}\%$	S_1, S_2, S_3

Because S_1 is contested among the choosers, set it aside. Now randomly give Jean S_2 or S_3. Assume she receives S_2. Now combine the remaining shares and use the divider-chooser method to distribute the remaining two shares. Randomly choose Caroline or Jasmine to be the divider. The other partner is the chooser. We will assume Caroline is the divider and divides the shares as T_1 and T_2. Assume Jasmine chooses T_1. Then Caroline gets T_2, Jasmine gets T_1, and Jean gets S_2.

9. Answers may vary. Here is one possibility.

	S_1	S_2	S_3	S_4	Bids
A	25%	25%	25%	25%	S_1, S_2, S_3, S_4
B	30%	10%	30%	30%	S_1, S_3, S_4
C	20%	35%	25%	20%	S_2, S_3
D	30%	30%	20%	20%	S_1, S_2

S_4 is uncontested among the choosers. Award S_4 to B. Combine S_1, S_2, and S_3. Then choose one of the remaining players to be the divider. We choose A. A divides the shares into three shares: T_1, T_2, and T_3. Everyone now submits bids for those shares. *This is where solutions may vary. It will depend on how you assign the bids. This is just one possibility.*

	T_1	T_2	T_3	Bids
A	$33\frac{1}{3}\%$	$33\frac{1}{3}\%$	$33\frac{1}{3}\%$	T_1, T_2, T_3
C	30%	30%	40%	T_3
D	30%	35%	35%	T_2, T_3

T_1 is uncontested, so award that to A. C only wants T_3 so award T_3 to C. That leaves D with T_2. The final result: A gets T_1; B gets S_4; C gets T_3; D gets T_2.

11. Answers may vary. Here is one possibility.

	S_1	S_2	S_3	S_4	Bids
A	25%	25%	25%	25%	S_1, S_2, S_3, S_4
R	15%	30%	20%	35%	S_2, S_4
M	20%	45%	15%	20%	S_2
C	30%	35%	15%	20%	S_1, S_2

Michel only values S_2, so he receives S_2. Because S_2 went to Michel, award S_4 to Richard. Then S_1 goes to Carlos. Ariana gets S_3.

13. Answers may vary. This is one possibility.

	S_1	S_2	S_3	S_4	Bids
J	25%	25%	25%	25%	S_1, S_2, S_3, S_4
Y	35%	20%	10%	35%	S_1, S_4
M	15%	40%	35%	10%	S_2, S_3
D	25%	35%	10%	30%	S_1, S_2, S_4

S_3 is uncontested among the choosers. Award S_3 to Jorge. Now choose one of the remaining investors to be the divider. We choose Yolanda. She divides the shares into three shares: T_1, T_2, and T_3. Everyone now submits bids for those shares. *This is where solutions may vary. It will depend on how you assign the bids. This is just one possibility.*

	T_1	T_2	T_3	Bids
Y	$33\frac{1}{3}\%$	$33\frac{1}{3}\%$	$33\frac{1}{3}\%$	T_1, T_2, T_3
M	30%	30%	40%	T_3
D	30%	35%	35%	T_2, T_3

T_2 is uncontested among the choosers. Award T_2 to Daniel. Now use the divider-chooser method for two people, Yolanda and Maria, to divide the remaining two shares: call them U_1 and U_2. We will assume Maria divides and Yolanda chooses U_2. The final result: Jorge gets S_3, Yolanda gets U_2, Maria gets U_1, and Daniel gets T_2.

15. Assign initial winners:

	Chong	Riana	Winner
Boat	15	5	Chong
House	45	65	Riana
Condo	40	30	Chong
Total:	55	65	

Riana is the initial winner and Chong is the initial loser.

Adjusted winner procedure:

	Chong	Riana	Winner	Ratio
Boat	15	5	Chong	
House	45	65	Riana	$\frac{65}{45} \approx 1.4$
Condo	40	30	Chong	
Total:	55	65		

The sharing procedure for the house:

$$15 + 45(1 - x) + 40 = 65x$$
$$15 + 45 - 45x + 40 = 65x$$
$$100 - 45x = 65x$$
$$100 = 110x$$
$$0.909 \approx x$$

For the house, 90.9% will go to Riana, and 9.1% will go to Chong.

Distribute the estate: Chong gets the boat, condo and 9.1% from the sale of the house. Riana gets 90.9% of the sale of the house.

17. Assign initial winners:

	Elliot	Francine	Winner
Residence	20	25	Francine
Summer Home	10	10	tie
Airplane	18	25	Francine
Sailboat	20	15	Elliot
Desk	32	25	Elliot
Total:	52	50	

Elliot is the initial winner and Francine is the initial loser.

The sharing procedure:

$$10(1 - x) + 20 + 32 = 25 + 10x + 25$$
$$10 - 10x + 20 + 32 = 25 + 10x + 25$$
$$62 - 10x = 50 + 10x$$
$$12 = 20x$$
$$0.60 = x$$

For the summer home, 60% will go to Francine, and 40% will go to Elliot.

Distribute the estate: Francine gets the residence, airplane, and 60% from the sale of the summer home. Elliot gets the sailboat, desk, and 40% of the sale of the summer home.

19. Determine highest bidder.

	M. Suites	Real Estate
Amelia	\$510,000	\$480,000
Justine	\$624,000	\$540,000
Colby	\$540,000	\$545,000
Highest bidder	Justine	Colby

Amelia: $\$510,000 + \$480,000 = \$990,000$

Fair share: $\frac{1}{3}(\$990,000) = \$330,000$

Justine: $\$624,000 + \$540,000 = \$1,164,000$

Fair share: $\frac{1}{3}(\$1,164,000) = \$388,000$

Colby: $\$540,000 + \$545,000 = \$1,085,000$

Fair share: $\frac{1}{3}(\$1,085,000) = \$361,667$

Difference between fair share and item won.
Amelia: \$330,000 – \$0 = \$330,000
Justine: \$388,000 – \$624,000 = –\$236,000
Colby: \$361,667 – \$545,000 = –\$183,333

Estate surplus:
–\$330,000 + \$236,000 + \$183,333 = \$89,333

Equal amount = $\frac{1}{3}(\$89,333) = \$29,777$

Amounts owed to/from estate:
Amelia: \$330,000 + \$29,777 = \$359,777
Justine: –\$236,000 + \$29,777 = –\$206,223
Colby: –\$183,333 + \$29,777 = –\$153,556

Amelia receives \$359,777 from the estate. Justine gets the medical suites and pays \$206,223 into the estate. Colby gets the real estate and pays \$153,556 into the estate.

21. Determine highest bidder. (in thousands)

	Leonard	Taylor	Madison	High
Home	\$600	\$520	\$475	Leonard
SUV	\$12	\$10	\$14	Madison
Boat	\$58	\$70	\$50	Taylor
Paint	\$110	\$120	\$121	Madison

Leonard: \$600 + \$12 + \$58 + \$110 = \$780

Fair share: $\frac{1}{3}(\$780) = \$260$

Taylor: \$520 + \$10 + \$70 + \$120 = \$720

Fair share: $\frac{1}{3}(\$720) = \$240$

Madison: $475 + $14 + $50 + $121 = $660

Fair share: $\frac{1}{3}(\$660) = \$220$

Difference between fair share and item won.
Leonard: $260 – $600 = –$340
Taylor: $240 – $70 = $170
Madison: $220 – $14 – $121 = $85

Estate surplus:
$340 – $170 – $85 = $85

Equal amount = $\frac{1}{3}(\$85) = \$28.333$

Amounts owed to/from estate:
Leonard: –$340 + $28.333 = –$311.667
Taylor: $170 + $28.333 = $198.333
Madison: $85 + $28.333 = $113.333

Leonard gets the home and pays $311,667 into the estate. Taylor gets the boat and receives $198,333 from the estate. Madison gets the SUV and painting and receives $113,333 from the estate.

23. Determine who gets the room.

	Clara	Manuel	Charles	Beatrice	Room to
1	$500	$800	$500	$600	Manuel
2	$600	$400	$500	$700	Beatrice
3	$750	$700	$900	$800	Charles
4	$650	$600	$600	$400	Clara

	Clara	Manuel	Charles	Beatrice
Total Valuation	$2500	$2500	$2500	$2500
Fair share	$625	$625	$625	$625
Value of bd won	$650	$800	$900	$700
Owed to rent	–$25	–$175	–$275	–$75
Receives from rent	0	0	0	0
Share of surplus	$137.5	$137.5	$137.5	$137.5
Disposition of money	$112.5	–$37.5	–$137.5	$62.5

Rent surplus:
$25 + $175 + $275 + $75 = $550

Equal amount = $\frac{1}{4}(\$550) = \$137.50$

Rent paid:
Clara: $625 – $112.50 = $512.50
Manuel: $625 + $37.50 = $662.50
Charles: $625 + $137.50 = $762.50
Beatrice: $625 – $62.50 = $562.50

Clara gets bedroom 4 and pays $512.50.
Manuel gets bedroom 1 and pays $662.50.
Charles gets bedroom 3 and pays $762.50.
Beatrice gets bedroom 2 and pays $562.50.

25. Determine winner:

	Lenore	Raven	Winner
Chalet	$500,000	$600,000	Raven

$\frac{n-1}{n} = \frac{2-1}{2} = \frac{1}{2}$

$\frac{1}{2}(600,000) = 300,000$

Escrow account contributions:

Lenore: $\frac{1}{2}(-500,000) = -250,000$

Raven: 300,000

Total in escrow: –250,000 + 300,000 = 50,000
Divide 50,000 by 2 = 25,000.

Final net amount:
Lenore: 250,000 + 25,000 = 275,000
Raven: chalet – 300,000 + 25,000
= chalet – 275,000

Verify the division is fair:

Lenore: $\frac{275,000}{500,000} = 0.55$

Raven: $\frac{325,000}{600,000} \approx 0.54$

The division is fair.

Raven receives chalet and pays $275,000.
Lenore receives $275,000.

27. Determine winner: Item 1 is sailboat. Item 2 is RV. Item 3 is Cabin.

	Marshall	Jordan	Jan	Winner
1	$75,000	$60,000	$63,000	Marshall
2	$66,000	$63,000	$57,000	Marshall
3	$87,000	$90,000	$75,000	Jordan

$\frac{n-1}{n} = \frac{3-1}{3} = \frac{2}{3}$

Item 1: sailboat

Escrow account contributions:

Marshall owes: $\frac{2}{3}(75,000) = 50,000$

Jordan receives: $\frac{1}{3}(60,000) = 20,000$

Jan receives: $\frac{1}{3}(63,000) = 21,000$

Money left in the joint account:
50,000 – 20,000 – 21,000 = 9000
Divide 9000 by 3 = 3000.

Final accounting for the sailboat:
Marshall: sailboat – 50,000 + 3000
= sailboat – $47,000
Jordan: 20,000 + 3000 = $23,000
Jan: 21,000 + 3000 = $24,000

Item 2: RV

Escrow account contributions:

Marshall owes: $\frac{2}{3}(66,000) = 44,000$

Jordan receives: $\frac{1}{3}(63,000) = 21,000$

Jan receives: $\frac{1}{3}(57,000) = 19,000$

Money left in the joint account:
$44{,}000 - 21{,}000 - 19{,}000 = 4000$
Divide 4000 by 3 = 1333.
Final accounting for the RV:
Marshall: $\text{RV} - 44{,}000 + 1333 = \text{RV} - \$42{,}667$
Jordan: $21{,}000 + 1333 = \$22{,}333$
Jan: $19{,}000 + 1333 = \$20{,}333$

Item 3: cabin
Escrow account contributions:

Marshall receives: $\frac{1}{3}(87{,}000) = 29{,}000$

Jordan owes: $\frac{2}{3}(90{,}000) = 60{,}000$

Jan receives: $\frac{1}{3}(75{,}000) = 25{,}000$

Money left in the joint account:
$-29{,}000 + 60{,}000 - 25{,}000 = 6000$
Divide 6000 by 3 = 2000.
Final accounting for the cabin:
Marshall: $29{,}000 + 2000 = 31{,}000$
Jordan: $\text{cabin} - 60{,}000 + 2000 = \text{cabin} - \$58{,}000$
Jan: $25{,}000 + 2000 = \$27{,}000$

Disposition of sailboat, RV, cabin, and money:
Marshall:
$\text{sailboat} - 47{,}000 + \text{RV} - 42{,}667 + 31{,}000 = \text{sailboat} + \text{RV} - \$58{,}667$
Jordan: $23{,}000 + 22{,}333 + \text{cabin} - 58{,}000 = \text{cabin} - \$12{,}667$
Jan: $24{,}000 + 20{,}333 + 27{,}000 = \$71{,}333$

Thus, Marshall receives the sailboat and RV and pays \$58,667. Jordan receives the cabin and pays \$12,667. Jan receives \$71,333.

29. Determine winner:
Item 1 is piano. Item 2 is organ. Item 3 is studio. Item 4 is condo. (Amounts in thousands)

	June	Melissa	Lloyd	Diane	Winner
1	\$15	\$10	\$12	\$10	June
2	\$9	\$12	\$10	\$9	Melissa
3	\$195	\$225	\$175	\$200	Melissa
4	\$175	\$250	\$225	\$200	Melissa

$$\frac{n-1}{n} = \frac{4-1}{4} = \frac{3}{4}$$

Item 1: piano
Escrow account contributions:

June owes: $\frac{3}{4}(15) = 11.25$

Melissa receives: $\frac{1}{4}(10) = 2.5$

Lloyd receives: $\frac{1}{4}(12) = 3$

Diane receives: $\frac{1}{4}(10) = 2.5$

Money left in the joint account:
$11.25 - 2.5 - 3 - 2.5 = 3.25$
Divide 3.25 by 4 = 0.8125
Final accounting for the piano:
June: $\text{piano} - 11.25 + 0.8125 = \text{piano} - \$10.4375$
Melissa: $2.5 + 0.8125 = \$3.3125$
Lloyd: $3 + 0.8125 = \$3.8125$
Diane: $2.5 + 0.8125 = \$3.3125$

Item 2: organ
Escrow account contributions:

June receives: $\frac{1}{4}(9) = 2.25$

Melissa owes: $\frac{3}{4}(12) = 9$

Lloyd receives: $\frac{1}{4}(10) = 2.5$

Diane receives: $\frac{1}{4}(9) = 2.25$

Money left in the joint account:
$-2.25 + 9 - 2.5 - 2.25 = 2$
Divide 2 by 4 = 0.5
Final accounting for the organ:
June: $2.25 + 0.5 = \$2.75$
Melissa: $\text{organ} - 9 + 0.5 = \text{organ} - \$8.5$
Lloyd: $2.5 + 0.5 = \$3$
Diane: $2.25 + 0.5 = \$2.75$

Item 3: studio
Escrow account contributions:

June receives: $\frac{1}{4}(195) = 48.75$

Melissa owes: $\frac{3}{4}(225) = 168.75$

Lloyd receives: $\frac{1}{4}(175) = 43.75$

Diane receives: $\frac{1}{4}(200) = 50$

Money left in the joint account:
$-48.75 + 168.75 - 43.75 - 50 = 26.25$
Divide 26.25 by 4 = 6.5625
Final accounting for the studio:
June: $48.75 + 6.5625 = \$55.3125$
Melissa: $\text{studio} - 168.75 + 6.5625 = \text{studio} - \$162.1875$
Lloyd: $43.75 + 6.5625 = \$50.3125$
Diane: $50 + 6.5625 = \$56.5625$

Item 4: condo
Escrow account contributions:

June receives: $\frac{1}{4}(175) = 43.75$

Melissa owes: $\frac{3}{4}(250) = 187.5$

Lloyd receives: $\frac{1}{4}(225) = 56.25$

Diane receives: $\frac{1}{4}(200) = 50$

Money left in the joint account:
$-43.75 + 187.5 - 56.25 - 50 = 37.5$
Divide 37.5 by 4 = 9.375

Final accounting for the condo:
June: 43.75 + 9.375 = \$53.125
Melissa: condo – 187.5 + 9.375
= studio – \$178.125
Lloyd: 56.25 + 9.375= \$65.625
Diane: 50 + 9.375= \$59.375

Disposition of piano, organ, studio, condo, and money:
June:
piano – 10.4375 + 2.75 +55.3125 + 53.125
= piano + \$100.75
Melissa: 3.3125 + organ – 8.5 + studio
– 162.1875 + condo – 178.125
= organ + studio + condo – \$345.5
Lloyd: 3.8125 + 3 + 50.3125 + 65.625
= \$122.75
Diane: 3.3125 + 2.75 + 56.5625 + 59.375
= \$122

Thus, June receives the piano and \$100,750. Melissa receives the organ, studio, and condo and pays \$345,500. Lloyd receives \$122,750. Diane receives \$122,000.

CHAPTER 1 REVIEW EXERCISES

1. $\frac{1}{32}$ Multiply by $\frac{1}{2}$.

2. –64 Multiply by –2.

3. –2. Subtract 3, then add 2, then subtract 3, then add 2, etc.

4. 64. Multiply by 4, then divide by 2, then multiply by 4, then divide by 2, etc.

5. 1000
This is counting with binary numbers.

6. 1 6 15 20 15 6 1
This is Pascal's triangle.
Start the row with 1, then add the two numbers above to get 6 (1+5), 15 (5+10), 20 (10+10), 15 (10+5), 6 (5+1), and then end with 1.

7. Each image has a new circle.

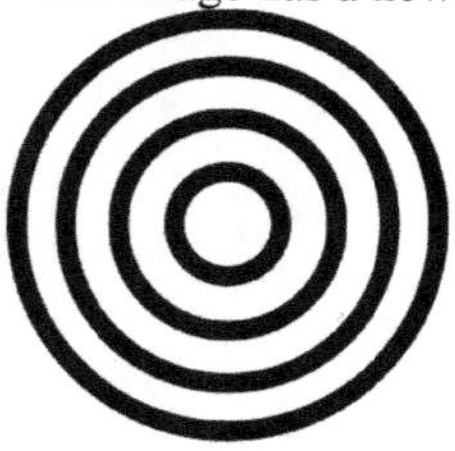

8. Each image has 2, 4, 8 gray rectangles. The next image will have 16 gray rectangles. Each image has one less white rectangle than gray rectangles: 1, 3, 7. The next image will have 15 white rectangles.

9. Each image has a 6-sided figure. Figures alternate gray interior and white interior. The square and triangles move in counter-clockwise position.

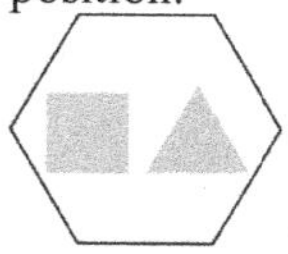

10. Each image has a square. The black interior square moves in clockwise direction. The gray interior circle moves in counter-clockwise direction.

11. More than. From the graph, it appears that as the Number of Units Willing to Produce increases, the Cost Per Unit increases.

12. 10 cases. From the bar graph, it appears that as the Months increases, the number of cases decreases by 3.

13. More than. From the table, the time (in seconds) is increasing by 0.5 second. The distance (in meters) 0^2, 2^2, 4^2, ... , 16^2. So at 4.5 seconds the distance is 18^2, or 324 meters, which is more than 316 meters.

14. Less than. From the table, as the Depth (in meters) increases, the Percent of Light decreases. Since the last entry is 160 meters and 9% light, it is expected that for 180 meters, the percent of light will be less than 9%.

15. Answers may vary. One possibility is 0.5.

16. 0 does not have a reciprocal.

17. Answers may vary. One possibility is –3(6) = –18.

18. Answers may vary. One possibility is rectangle 1 has sides 10, 40, 10, and 40. The perimeter is 100 units; the area is 400 square units. Rectangle 2 has sides 20, 30, 20, and 30. The perimeter is 100 units; the area is 600 square units.

19.

	CS	Chem	Bus	Bio
M	X1	X3	X3	✓
C	X4	X2	✓	X2
R	✓	X2	X4	X2
E	X3	✓	X3	X3

Michael is a Biology major; Clarissa is a Business major; Reggie is a Computer Science major; Ellen is a Chemistry major.

20.

	Bank	Super	Service	Drug
D	X4	X3	X3	✓
P	X2	✓	X4	X2
T	✓	X3	X1	X4
G	X2	X3	✓	X2

Dodgers sponsored by drugstore; Pirates sponsored by supermarket; Tigers sponsored by bank; Giants sponsored by service station.

21. First weighing: Place 5 coins at each side of the scale. Exclude all the 5 coins in the side that goes up (the heavier coin is on the pan that went down). Now you have only 5 coins.
Second Weighing: Hold one coin in your hand and put 2 coins at each side of the scale. If the two sides balance, then the heavier one is the one you are holding in your hand. You are done. If one side is heavier than the other, one of the two coins is on that side is the heavier one. Now you have only 2 coins and a scale to find out which is the heavier by using the scale for a third time.

22. $500 \text{ bricks} \times \frac{5 \text{ pounds}}{1 \text{ brick}} = 2500 \text{ pounds}$
No, since 2500 > 2000 the truck cannot haul the pallet of bricks.

23. $1 \text{ inch} > 2.5 \text{ cm}$
$10(1 \text{ inch}) > 10(2.5 \text{ cm})$
$10 \text{ inches} > 25 \text{ cm}$
More than

24. $\frac{3000 \text{ words}}{300 \text{ words/minute}} = 10 \text{ minutes}$

25. Answers will vary depending on the grid. Using a 5 by 5 grid, we count 23 berries. Multiply 23 by the number of sections, 25, to get 575. So our estimate is 575 berries.

26. $72 \text{ inches} \cdot \frac{1}{8} = 9 \text{ inches}$

27. $41{,}320{,}000{,}000{,}000 = 4.132 \times 10^{13}$

28. $0.00000000012 = 1.2 \times 10^{-10}$

29. The difference, 7, is greatest between 2015 and 2016.

30. Physicians, since that is the largest section of the chart.

31. a. The largest decrease is between Years 0 and 1.

b. The end of Year 4 is the beginning of Year 5, which is about $14,000. The car's original value is $35,000. The car's value is less than one-half of the original value.

32. If 11 is across from 35, then there are
$(35 - 11) - 1 = 23$ people
on each side of person 11 and person 35. So the total number of people is 2(23) + 2 = 48 people.

33. Seth and Sarah cross the bridge, and Sarah returns. Time for round trip, 3 minutes.
Kevin and Asher cross the bridge, and Seth returns. Time for round trip is 12 minutes. Total elapsed time is 15 minutes.
Now Sarah and Seth cross the bridge. Total elapsed time is 17 minutes.

34. Start with 9 red balls and 9 blue balls. Because there is an odd number of each color of ball, the last two balls selected from the box will be different colors, resulting in one red ball in the box.

35. No. A fair share is $6. If piece 1 is one-third pepperoni and one-sixth cheese, then the value of the piece is
$$\frac{1}{3}(9) + \frac{1}{6}(3) = 3 + 2 = 5 < 6.$$

36. Answers may vary. Here is one possibility.

	S_1	S_2	S_3	S_4	Bids
J	25%	25%	25%	25%	S_1, S_2, S_3, S_4
Cl	15%	35%	20%	30%	S_2, S_4
A	15%	45%	20%	20%	S_2
Ca	20%	35%	10%	35%	S_2, S_4

Jeffrey only values S_3, so he receives S_3.
Now combine the remaining shares and randomly select a new divider. Suppose Clara is chosen as the divider and separates the assets into shares that to her are equal. Alyssa and Cameron bid on the shares as shown in the following table. *This is where solutions may vary. It will depend on how you assign the bids. This is just one possibility.*

	T_1	T_2	T_3	Bids
Cl	$33\frac{1}{3}\%$	$33\frac{1}{3}\%$	$33\frac{1}{3}\%$	T_1, T_2, T_3
A	25%	40%	35%	T_2, T_3
Ca	20%	35%	45%	T_2, T_3

Because T_2 and T_3 are contested and T_1 is uncontested by the choosers, award T_1 to Clara. Now use the divider-chooser method for two people, Alyssa and Cameron, to divide the remaining two shares; call them U_1 and U_2. We will assume Alyssa divides and Cameron chooses U_2. The final result: Cameron gets U_2, Alyssa gets U_1, Clara gets T_1, and Jeffrey gets S_3.

37. Assign initial winners:

	Alana	Taft	Winner
Condo	40	50	Taft
Motorcycle	10	5	Alana
Piano	25	20	Alana
Painting	25	25	tie
Total:	35	50	

Taft is the initial winner and Alana is the initial loser.

The sharing procedure:

$$50 + 25(1 - x) = 10 + 25 + 25x$$
$$50 + 25 - 25x = 10 + 25 + 25x$$
$$75 - 25x = 35 + 25x$$
$$40 = 50x$$
$$0.80 = x$$

For the painting, 80% will go to Alana, and 20% will go to Taft.

Distribute the estate: Alana gets the motorcycle, piano, and 80% from the sale of the painting. Taft gets the condo and 20% of the sale of the painting.

38. Determine highest bidder.

	House	Condo
Miguel	\$550,000	\$300,000
Cannon	\$450,000	\$350,000
Ralph	\$350,000	\$400,000
Highest bidder	Miguel	Ralph

Miguel: $\$550,000 + \$300,000 = \$850,000$

Fair share: $\frac{1}{3}(\$850,000) \approx \$283,333$

Cannon: $\$450,000 + \$350,000 = \$800,000$

Fair share: $\frac{1}{3}(\$800,000) = \$266,667$

Ralph: $\$350,000 + \$400,000 = \$750,000$

Fair share: $\frac{1}{3}(\$750,000) = \$250,000$

Difference between fair share and item won.
Miguel: \$283,333 – \$550,000 = –\$266,667
Cannon: \$266,667 – \$0 = \$266,667
Ralph: \$250,000 – \$400,000 = –\$150,000

Estate surplus:
\$266,667 – \$266,667 + \$150,000 = \$150,000

Equal amount = $\frac{1}{3}(\$150,000) = \$50,000$

Amounts owed to/from estate:
Miguel: –\$266,667 + \$50,000 = –\$216,667
Cannon: \$266,667 + \$50,000 = \$316,667
Ralph: –\$150,000 + \$50,000 = –\$100,000

Miguel gets the house and pays \$216,667 into the estate. Cannon receives \$316,667 from the estate. Ralph gets the condo and pays \$100,000 into the estate.

39. Determine who gets the room.

	Thomas	Michael	Claudia	Room to
Garden	\$1000	\$1100	\$1000	Michael
Skyline	\$900	\$900	\$1000	Claudia
No view	\$800	\$700	\$700	Thomas

	Thomas	Michael	Claudia
Total Valuation	\$2700	\$2700	\$2700
Fair share	\$900	\$900	\$900
Value of bd won	\$800	\$1100	\$1000
Owed to rent		–\$200	–\$100
Receives from rent	\$100	0	0
Share of surplus	\$67	\$67	\$67
Disposition of money	\$167	–\$133	–\$33

Rent surplus:
–\$100 + \$200 + \$100 = \$200

Equal amount = $\frac{1}{3}(\$200) = \$67$

Rent paid:
Thomas: \$900 – \$167 = \$733
Michael: \$900 + \$133 = \$1033
Claudia: \$900 + \$33 = \$933

Thomas gets the no view and pays \$733. Michael gets the garden view and pays \$1033. Claudia gets the skyline view and pays \$933.

40. Determine winner:
Item 1 is vase. Item 2 is violin. Item 3 is painting. (Amounts in thousands)

	Jocelyn	Harrison	Colby	Winner
1	\$810	\$900	\$750	Harrison
2	\$1200	\$1500	\$1200	Harrison
3	\$1500	\$1650	\$1800	Colby

$$\frac{n-1}{n}=\frac{3-1}{3}=\frac{2}{3}$$

Item 1: vase
Escrow account contributions:

Jocelyn receives: $\frac{1}{3}(810)=270$

Harrison owes: $\frac{2}{3}(900)=600$

Colby receives: $\frac{1}{3}(750)=250$

Money left in the joint account:
–270 + 600 – 250 = 80
Divide 80 by 3 = 26.666
Final accounting for the vase:
Jocelyn: 270 + 26.666 = \$296.666
Harrison: vase – 600 + 26.666
= vase – \$573.334
Colby: 250 + 26.666 = \$276.666

Item 2: violin
Escrow account contributions:

Jocelyn receives: $\frac{1}{3}(1200)=400$

Harrison owes: $\frac{2}{3}(1500)=1000$

Colby receives: $\frac{1}{3}(1200)=400$

Money left in the joint account:
–400 + 1000 – 400 = 200
Divide 200 by 3 = 66.666.
Final accounting for the violin:
Jocelyn: 400 + 66.666 = \$466.666
Harrison: violin – 1000 + 66.666
= violin –\$933.334
Colby: 400 + 66.666 = \$466.666

Item 3: painting
Escrow account contributions:

Jocelyn receives: $\frac{1}{3}(1500)=500$

Harrison receives: $\frac{1}{3}(1650)=550$

Colby owes: $\frac{2}{3}(1800)=1200$

Money left in the joint account:
–500 – 550 + 1200 = 150
Divide 150 by 3 = 50.
Final accounting for the painting:
Jocelyn: 500 + 50 = \$550
Harrison: 550 + 50 = \$600
Colby: painting – 1200 + 50
= painting – \$1150

Disposition of vase, violin, painting, and money:
Jocelyn: 296.666 + 466.666 + 550
= \$1313.332
Harrison:
vase – 573.334 + violin – 933.334 + 600
= vase + violin – \$906.668
Colby:
276.666 + 466.666 + painting – 1150
= painting – \$406.668

Thus, Jocelyn receives \$1,313,332. Harrison receives the vase and violin and pays \$906,668. Colby receives the painting and pays \$406,668.

41. Determine winner:
Item 1 is Bentley. Item 2 is condo. Item 3 is harp. Item 4 is sculpture. (Amounts in thousands)

	Santoro	Ben	Bianca	Vivian	Winner
1	\$150	\$125	\$120	\$130	Santoro
2	\$1000	\$1200	\$1500	\$1300	Bianca
3	\$12	\$12	\$8	\$16	Vivian
4	\$1500	\$2000	\$1800	\$1600	Ben

$$\frac{n-1}{n}=\frac{4-1}{4}=\frac{3}{4}$$

Item 1: Bentley
Escrow account contributions:

Santoro owes: $\frac{3}{4}(150)=112.5$

Ben receives: $\frac{1}{4}(125)=31.25$

Bianca receives: $\frac{1}{4}(120)=30$

Vivian receives: $\frac{1}{4}(130)=32.5$

Money left in the joint account:
112.5 – 31.25 – 30 – 32.5 = 18.75
Divide 18.75 by 4 = 4.6875
Final accounting for the Bentley:
Santoro: Bentley – 112.5 + 4.6875
= Bentley – \$107.8125
Ben: 31.25 + 4.6875 = \$35.9375
Bianca: 30 + 4.6875 = \$34.6875
Vivian: 32.5 + 4.6875 = \$37.1875

Item 2: condo
Escrow account contributions:

Santoro receives: $\frac{1}{4}(1000)=250$

Ben receives: $\frac{1}{4}(1200)=300$

Bianca owes: $\frac{3}{4}(1500)=1125$

Vivian receives: $\frac{1}{4}(1300)=325$

Money left in the joint account:
–250 – 300 +1125 – 325 = 250
Divide 250 by 4 = 62.5
Final accounting for the condo:
Santoro: 250 + 62.5 = \$312.5

Ben: 300 + 62.5 = \$362.5
Bianca: condo – 1125 + 62.5
= condo – \$1062.5
Vivian: 325 + 62.5 = \$387.5

Item 3: harp
Escrow account contributions:

Santoro receives: $\frac{1}{4}(12) = 3$

Ben receives: $\frac{1}{4}(12) = 3$

Bianca receives: $\frac{1}{4}(8) = 2$

Vivian owes: $\frac{3}{4}(16) = 12$

Money left in the joint account:
$-3 - 3 - 2 + 12 = 4$
Divide 4 by 4 = 1
Final accounting for the harp:
Santoro: 3 + 1 = \$4
Ben: 3 + 1 = \$4
Bianca: 2 + 1 = \$3
Vivian: harp – 12 + 1 = harp – \$11

Item 4: sculpture
Escrow account contributions:

Santoro receives: $\frac{1}{4}(1500) = 375$

Ben owes: $\frac{3}{4}(2000) = 1500$

Bianca receives: $\frac{1}{4}(1800) = 450$

Vivian receives: $\frac{1}{4}(1600) = 400$

Money left in the joint account:
$-375 + 1500 - 450 - 400 = 275$
Divide 275 by 4 = 68.75
Final accounting for the sculpture:
Santoro: 375 + 68.75 = \$443.75
Ben: sculpture – 1500 + 68.75
= sculpture – \$1431.25
Bianca: 450 + 68.75 = \$518.75
Vivian: 400 + 68.75 = \$468.75

Disposition of Bentley, condo, harp, sculpture, and money:
Santoro: Bentley–107.8125+312.5 +4+443.75
= Bentley + \$652.4375
Ben: 35.9375 +362.5 +4 +sculpture –1431.25
= sculpture – \$1028.8125
Bianca:
34.6875 + condo – 1062.5 + 3 + 518.75
= condo – \$506.0625
Vivian:
37.1875 + 387.5 + harp – 11 + 468.75
= harp + \$882.4375

Thus, Santoro receives the Bentley and \$652,437.50. Ben receives the sculpture and pays \$1,028,812.50. Bianca receives the condo and pays \$506,062.50. Vivian receives the harp and receives \$882,437.50.

Chapter 2: Sets and Logic

THINK ABOUT IT, SECTION 2.1

1. a. True
   b. True
   c. True
   d. False
   e. False
   f. True
   g. True
   h. True

2. a. No
   b. Yes

3. A and B are disjoint—that is, they have no elements in common.

4. $(A')' = A$

5. False negative

6. Correct outcome

EXERCISE SET 2.1

1. {penny, nickel, dime, quarter, 50 cent piece}

3. {Mercury, Mars}

5. {0, 1, 2, 3, 4, 5, 6, 7}

7. Solving:
   $2x - 1 = -11$
   $2x = -10$
   $x = -5$
   So {–5} is the solution set.

9. Because b is an element of the given set, the statement is true.

11. True

13. True

15. False; although {5} $\{5\} \subset \{1, 2, 3, 4, 5\}$.

17. The complement of {2, 4, 6, 7} contains elements in U but not in the set: {0, 1, 3, 5, 8}.

19. $\varnothing' = U = \{0, 1, 2, 3, 4, 5, 6, 7, 8\}$

21. {prime numbers}′ = {0, 1, 4, 6, 8}

23. Ø, {a}, {z}, {a, z}

25. Ø, {I}, {II}, {III}, {IV}, {I, II}, {I, III}, {I, IV}, {II, III}, {II, IV}, {III, IV}, {I, II, III}, {I, II, IV}, {I, III, IV}, {II, III, IV}, {I, II, III, IV}

27. Since there are 5 elements in the set {a, e, i, o, u}, there are $2^5 = 32$ subsets.

29. $2^{12} = 4{,}096$ different versions

31. $A \cup B = \{2, 4, 6\} \cup \{1, 2, 5, 8\}$
    $= \{1, 2, 4, 5, 6, 8\}$

33. $A \cap B' = \{2, 4, 6\} \cap \{3, 4, 6, 7\} = \{4, 6\}$

35. $(A \cup B)' = (\{1, 2, 4, 5, 6, 8\})' = \{3, 7\}$

37. $B \cap C = \{1, 2, 5, 8\} \cap \{1, 3, 7\} = \{1\}$.
    $A \cap (B \cap C) = \{2, 4, 6\} \cap \{1\} = \varnothing$

39. True

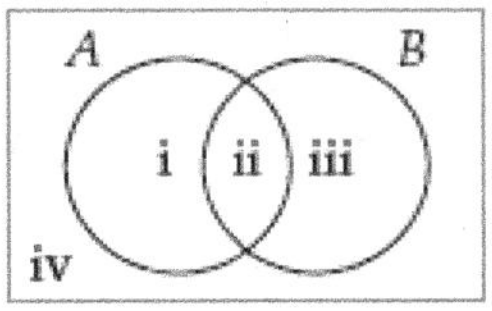

$A \cap B$: ii
Since ii $\subseteq A$, then $A \cap B \subseteq A$ is true.

41. True
    Since $A \subseteq U$, then $A \cap U = A$ is true.

43. $A' \cup B$

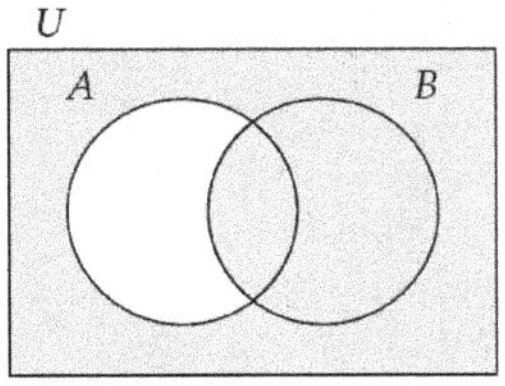

45. $A' \cup B'$

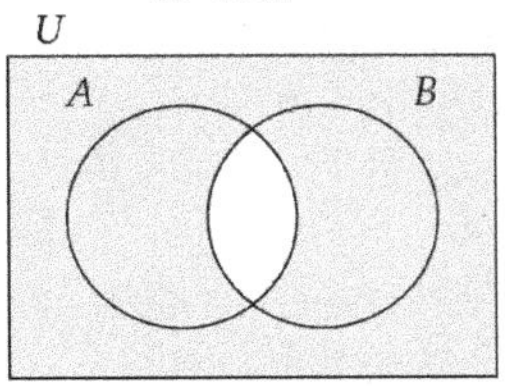

47. $(A \cap B)'$

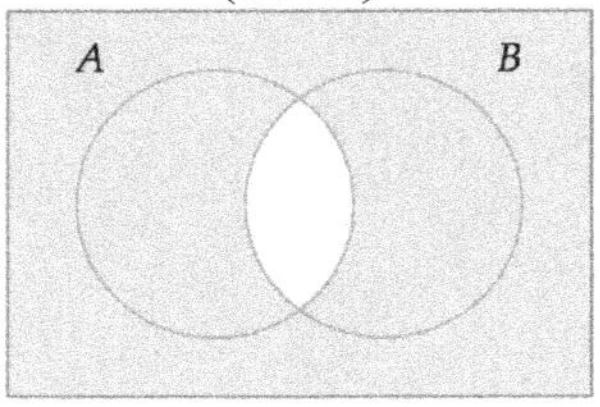

$A' \cup B'$

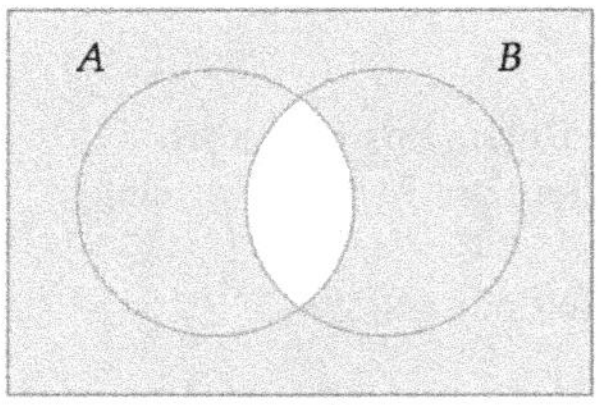

Therefore, $(A \cap B)' = A' \cup B'$.

49. $A \cap (B \cap C)$

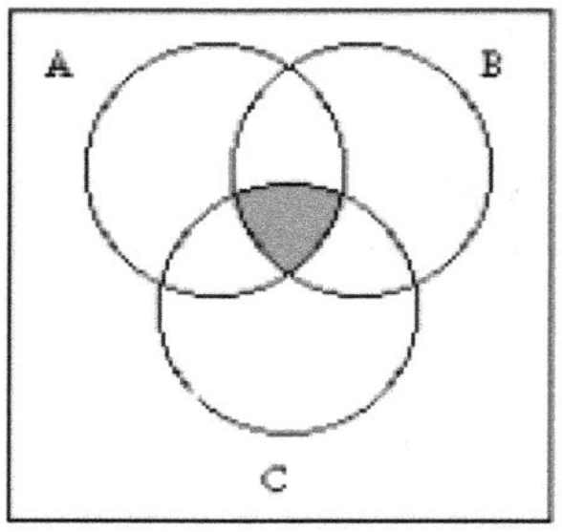

$(A \cap B) \cap C$

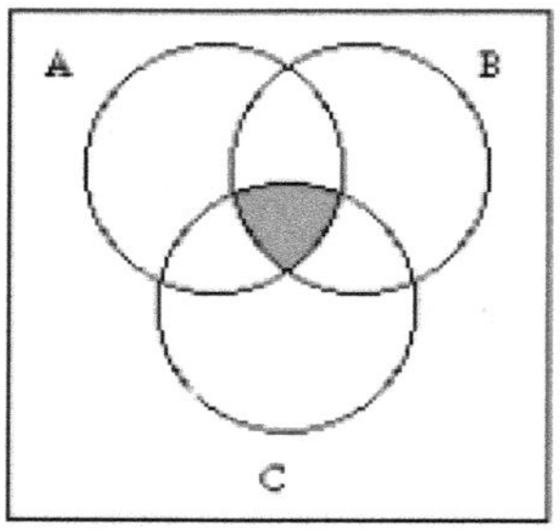

Therefore, $A \cap (B \cap C) = (A \cap B) \cap C$.

51. a. $S = \{\text{investors in stocks}\}$ and let $B = \{\text{investors in bonds}\}$. Since 75 had not invested in either stocks or bonds, $n(S \cup B) = 600 - 75 = 525$.

$$n(S \cup B) = n(S) + n(B) - n(S \cap B)$$
$$525 = 380 + 325 - n(S \cap B)$$
$$525 = 705 - n(S \cap B)$$
$$-180 = -n(S \cap B)$$
$$n(S \cap B) = 180$$

$n(S \cap B) = 180$ represents the number of investors in both stocks and bonds.

b. $n(S \text{ only}) = 380 - 180 = 200$

53. a. $1785 + 1219 + 831 = 3835$ people

b. 245 people

c. $755 + 700 + 275 = 1730$ people

d. $1785+755+700+245+1219+275+831=5810$
$6000 - 5810 = 190$ people

55. a. 101 people

b. $124 + 82 + 65 + 51 + 48 = 370$ people

c. $124 + 82 + 133 + 41 = 380$ people

d. $124 + 82 + 101 + 66 = 373$ people

e. $124 + 82 + 101 + 66 = 373$ people

f. $124 + 82 + 65 + 101 + 66 + 51 + 41$
$= 530$ people

57. a.

	File Is Infected	File Is Not Infected
Found	813	15
Did Not Find	27	42

b. 15

c. The virus software did not find 27 files that were infected.

59. a. 5

b. There were 5 cases in which the test was positive but the child did not have the genetic marker.

c. 12

d. There were 12 cases in which the test was negative but the child had the genetic marker.

e. 195

f. There were 195 cases in which the test was positive and the child had the genetic marker.

g. 288

h. There were 288 cases in which the test was negative and the child did not have the genetic marker.

i. Venn diagram

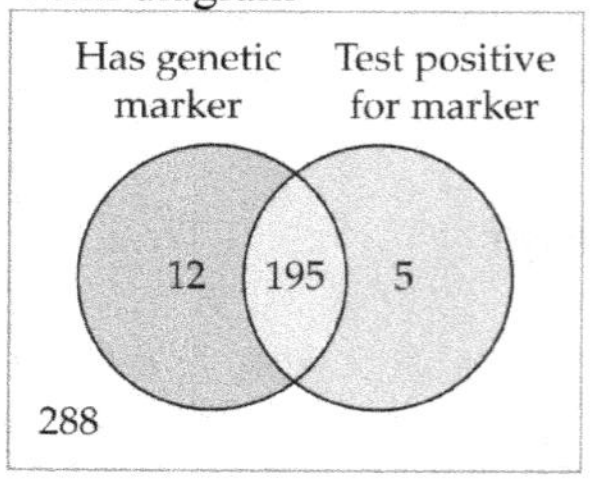

Think About It, Section 2.2

1. All humans are mammals. No cat is a dog.
2. Some pets are small. Some pets are not dogs.
3. No. The negation is: "Some Ferraris are not red."
4. Yes. For instance, "If I like ice cream, all numbers are positive. I like ice cream. Therefore, all numbers are positive." is a valid argument but the conclusion is false.

Exercise Set 2.2

1. No lions are playful.
3. Some classic movies were not first produced in black and white.
5. Some even numbers are odd numbers.
7. Some cars do not run on gasoline.
9. The argument is valid.

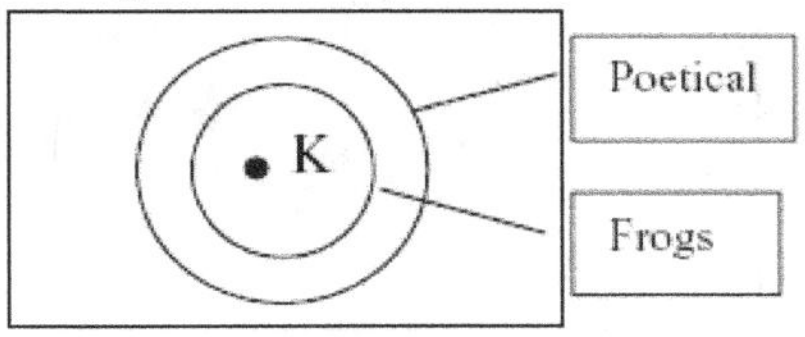

11. The argument is valid.

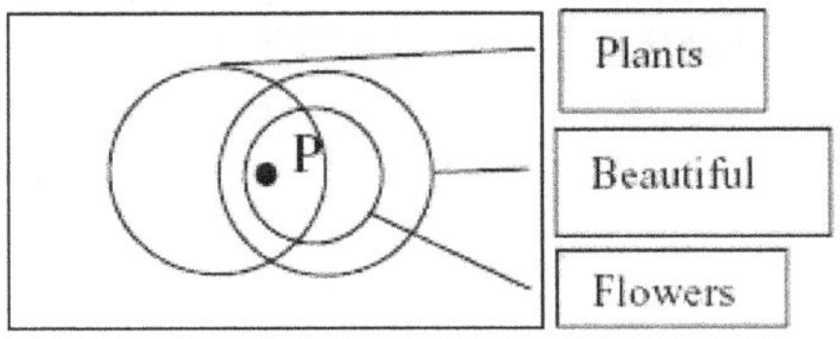

13. The argument is valid.

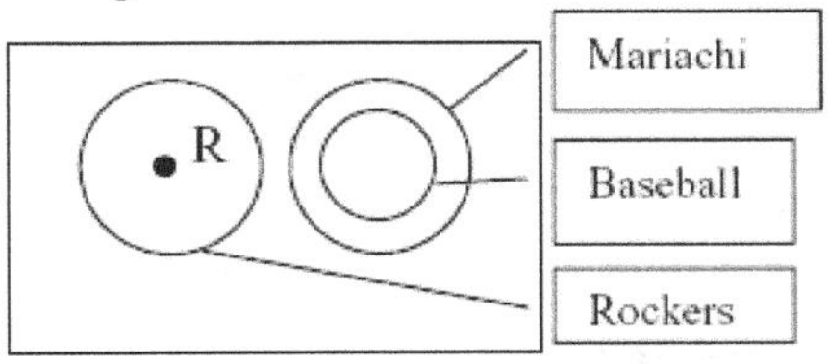

15. The argument is valid.

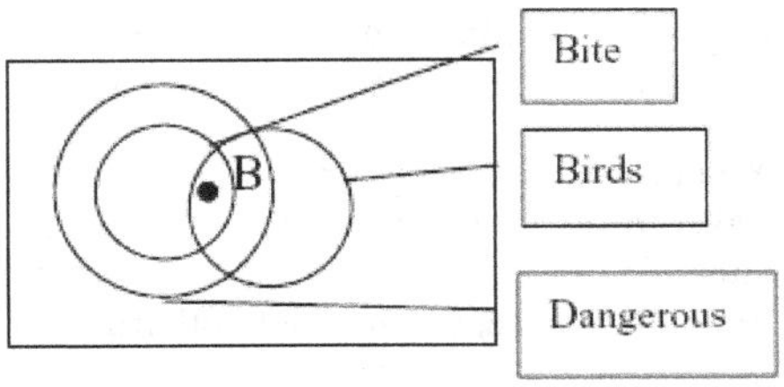

17. The argument is invalid.

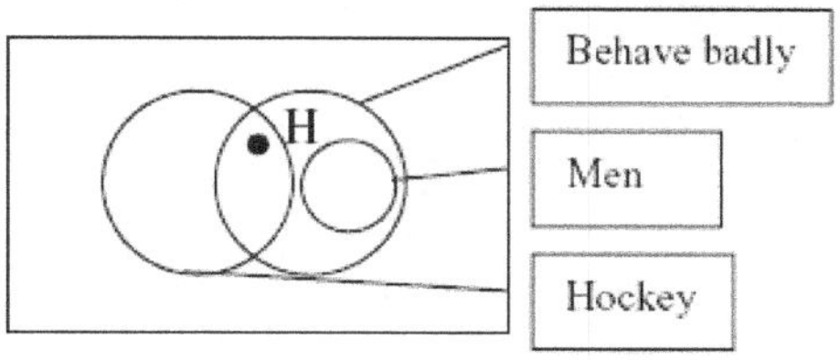

19. The argument is invalid.

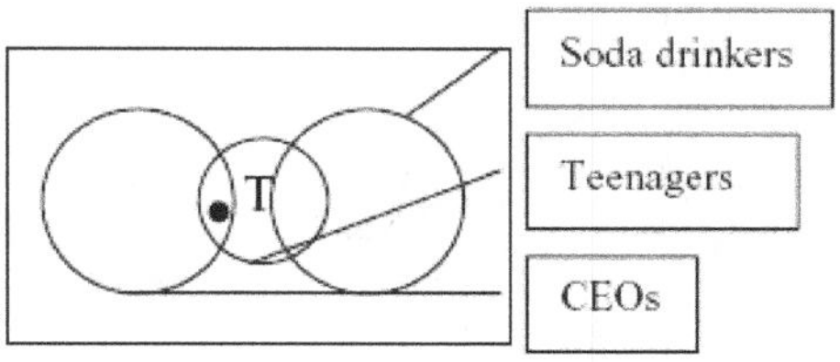

21. The argument is invalid.

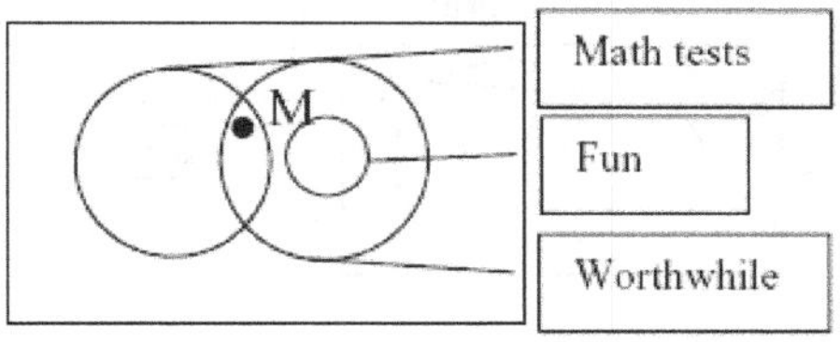

23. The argument is valid.

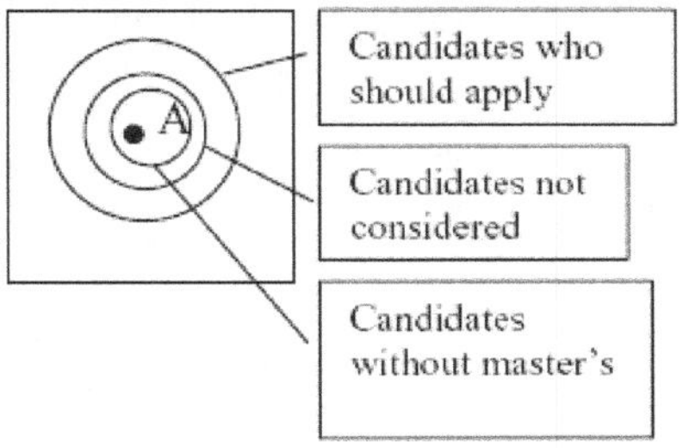

25. The argument is valid.

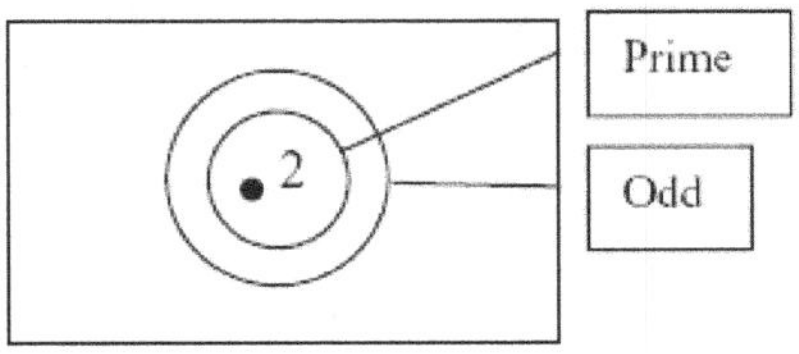

27. The argument is invalid.

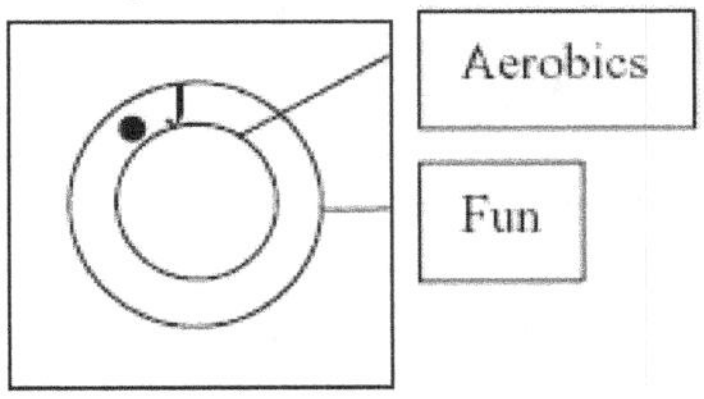

29. All Reuben sandwiches need mustard.

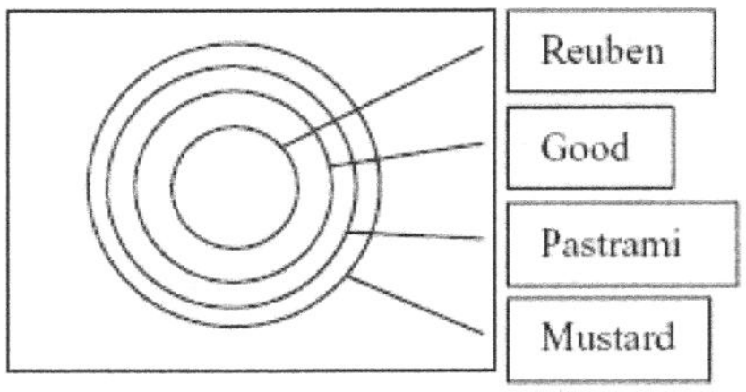

31. 1001 ends with a 5.

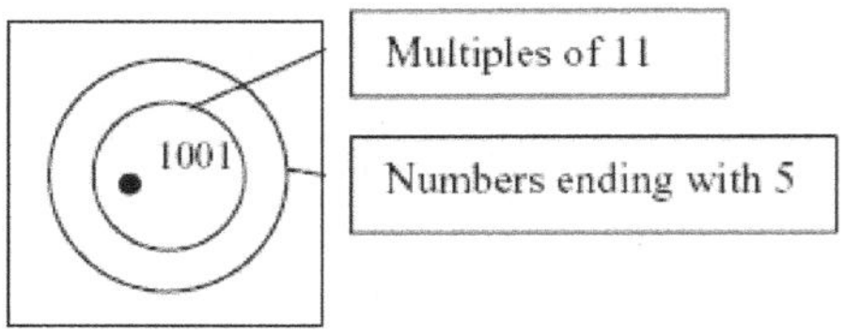

33. Some horses are gray.

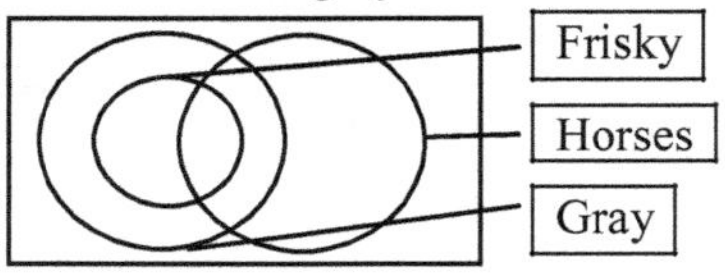

35. a. Invalid

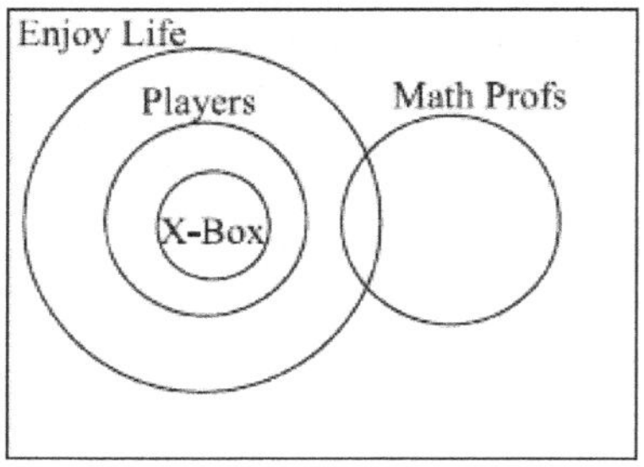

b. Invalid

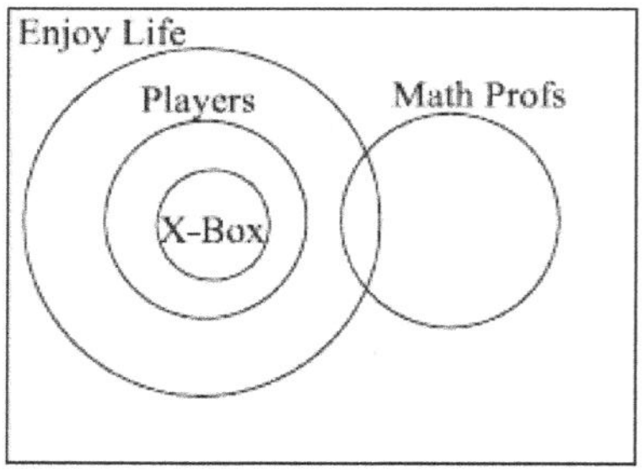

c. Invalid

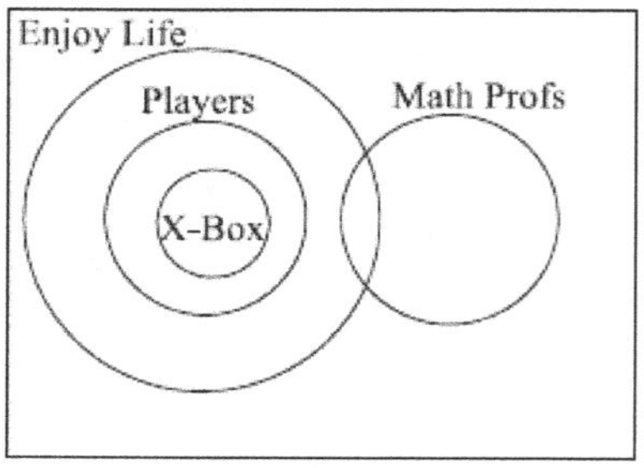

d. Invalid

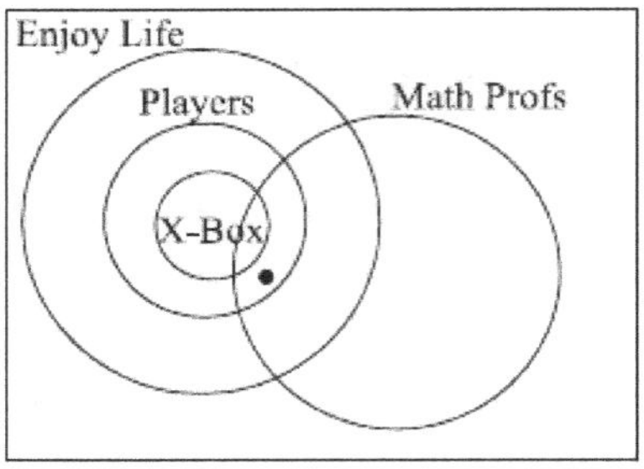

e. Valid

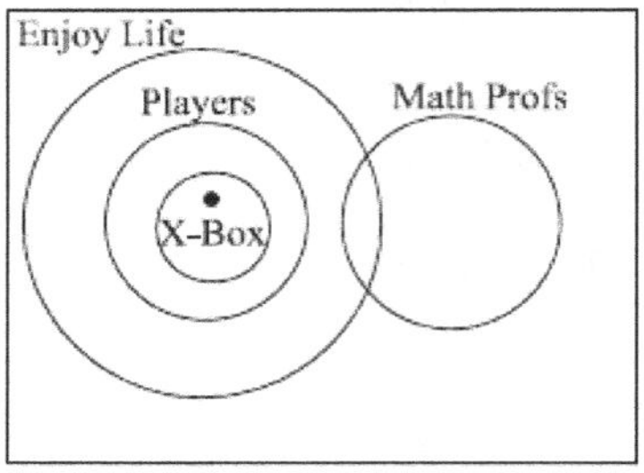

f. Valid

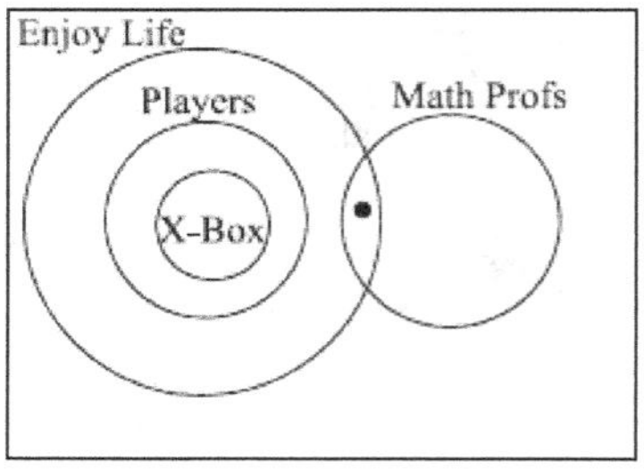

THINK ABOUT IT, SECTION 2.3

1. Or
2. And
3. If–then
4. True
5. False
6. True

EXERCISE SET 2.3

1. The sentence is an opinion. It can be true for one person and false for another person. Since it can be both true and false, it is not a statement.

3. You may not know the area code for Storm Lake, Iowa; however, you do know that it is either 512 or it is not 512. The sentence is either true or it is false, and it is not both true and false, so it is a statement.

5. The sentence "Have a fun trip," is a command, so it is not a statement.

7. The Giants did not lose the game.

9. The game did go into overtime.

11. $w \rightarrow t$

13. $d \rightarrow f$

15. $m \vee c$

17. I can play the piano and the violin.

19. If the painting is a watercolor, then James will not buy it.

21. If Carmella gets an A in mathematics, then she will major in physics or astronomy.

23. Let p represent the statement "It rained." Let q represent the statement "It snowed." In symbolic form, the original sentence is $\sim(p \vee q)$. One of De Morgan's laws states that this is equivalent to $\sim p \wedge \sim q$. Thus an equivalent sentence is "It did not rain and it did not snow."

25. Let p represent the statement "She visited France." Let q represent the statement "She visited Italy." In symbolic form, the original sentence is $\sim p \wedge \sim q$. One of De Morgan's laws states that this is equivalent to $\sim(p \vee q)$. Thus an equivalent sentence is "She did not visit either France or Italy."

27. *Antecedent:* I had the money.
Consequent: I would buy the painting.

29. *Antecedent:* She would go to the movies.
Consequent: Her friend would go.

31. *Antecedent:* Cameron gets a raise.
Consequent: She will buy a new tablet computer.

33. False. The antecedent is true and the consequent is false.

35. True. A false antecedent yields a true statement whether the consequent is true or false.

37. False. The antecedent is true and the consequent is false.

39. Converse: If I quit this job, then I would be rich.
Inverse: If I were not rich, then I would not quit this job.
Contrapositive: If I did not quit this job, I would not be rich.

41. Converse: If we are not able to attend the party, she will not return soon.
Inverse: If she returns soon, then we will be able to attend the party.
Contrapositive: If we are able to attend the party, she will return soon.

43. Converse: If we will take the train, we will be able to take the entire family.
Inverse: If we do not take the entire family, we can't take the train.
Contrapositive: If we do not take the train, we can't take the entire family.

45. In symbolic form:

$$\begin{array}{l} h \rightarrow r \\ \underline{\sim h} \\ \therefore \sim r \end{array}$$

h	r	**First premise** $h \rightarrow r$	**Second premise** $\sim h$	**Conclusion** $\sim r$
T	T	T	F	F
T	F	F	F	T
F	T	T	T	F
F	F	T	T	T

The premises are true in rows 3 and 4. Because the conclusion in row 3 is false and the premises are both true, we know the argument is invalid.

47. In symbolic form:

$$\begin{array}{l} e \rightarrow \sim s \\ \underline{\sim s} \\ \therefore e \end{array}$$

e	s	**First premise** $e \rightarrow \sim s$	**Second premise** $\sim s$	**Conclusion** e
T	T	F	F	T
T	F	T	T	T
F	T	T	F	F
F	F	T	T	F

The premises are true in rows 2 and 4. Because the conclusion in row 4 is false and the premises are both true, we know the argument is invalid.

49. Label the statements
f: You take Art 151 in the fall.
s: You will be eligible to take Art 152 in the spring.
In symbolic form:

$$\begin{array}{l} f \rightarrow s \\ \underline{\sim s} \\ \therefore \sim f \end{array}$$

The argument is valid using modus tollens.

51. Label the statements
    d: It is a dog.
    f: It has fleas.
    In symbolic form the argument is:
    In symbolic form:
    $d \to f$
    f
    $\therefore d$
    The argument is invalid using the fallacy of the converse.

53. Label the statements
    s: We serve salmon.
    v: Vicky will join us for lunch.
    m: Marilyn will join us for lunch.
    In symbolic form:
    $s \to v$
    $v \to \sim m$
    $\therefore s \to \sim m$
    The argument is valid using transitive reasoning.

55. In symbolic form the argument is

$s \to \sim r$	Premise 1
$\sim r \to c$	Premise 2
$\sim t \to \sim c$	Premise 3
$\therefore s \to t$	Conclusion

Applying transitive reasoning to Premise 1 and Premise 2 produces:

$s \to \sim r$	Premise 1
$\sim r \to c$	Premise 2
$\therefore s \to c$	

Premise 3 can be written as $c \to t$ using the contrapositive form. Combining Premise 3 with the conclusion from above gives

$s \to c$	Conclusion from above
$c \to t$	Equivalent form of Premise 3
$\therefore s \to t$	

This sequence of valid arguments has produced the desired conclusion. Thus the original argument is valid.

57. In symbolic form the argument is

$\sim o \to \sim f$	Premise 1
$c \to \sim o$	Premise 2
$\therefore f \to \sim c$	Conclusion

Premise 1 can be written as $f \to o$ using the contrapositive form. Premise 2 can be written as $o \to \sim c$ using the contrapositive form.
Applying the equivalent forms and transitive reasoning to Premise 1 and Premise 2 produces

$f \to o$	Equivalent form of Premise 1
$o \to \sim c$	Equivalent form of Premise 2
$\therefore f \to \sim c$	

This sequence of valid arguments has produced the desired conclusion. Thus the original argument is valid.

59. Label the statements
    m: The moon is made of green cheese.
    a: Apples are bananas.
    In symbolic form the argument is:

$m \to a$	Premise 1
$\sim a$	Premise 2
$\therefore ?$	

Applying modus tollens gives

$m \to a$	Premise 1
$\sim a$	Premise 2
$\therefore \sim m$	Modus tollens

Therefore, the moon is not made of green cheese.

61. Label the statements
    f: I pass the final exam.
    p: I will get a passing grade in the course.
    In symbolic form the argument is:

$p \to f$	Premise 1
f	Premise 2
$\therefore ?$	

The tempting conclusion is p, but this is the fallacy of the converse.

Therefore, a valid conclusion is not possible.

63. Label the statements
    t: It is a theropod.
    h: It is a herbivorous.
    s: It is a sauropod.
    In symbolic form the argument is:

$t \to \sim h$	Premise 1
$\sim h \to \sim s$	Premise 2
s	Premise 3
$\therefore ?$	

Applying transitive reasoning to Premise 1 and Premise 2 gives

$t \to \sim h$	Premise 1
$\sim h \to \sim s$	Premise 2
$\therefore t \to \sim s$	

Write the conclusion from above in its

equivalent contrapositive form, $s \rightarrow \sim t$. Use the equivalent form of the conclusion from above and Premise 3 to give

$s \rightarrow \sim t$ Equivalent form of conclusion
s Premise 3
$\therefore \sim t$ Modus ponens

Therefore, it is not a theropod.

THINK ABOUT IT, SECTION 2.4

1. Appeal to popularity fallacy.
2. Appeal to common practice fallacy.
3. Loaded question fallacy.

EXERCISE SET 2.4

1. Ad hominem
Premise: He doesn't believe in equal rights.
Conclusion: Don't listen to him.
This is an ad hominem fallacy because the premise is an attack on someone's character rather than the issue.

3. Black or white
Premise: Either you buy me a car or I will lose my job.
Conclusion: You must buy me a car so I don't lose my job.
This is a black-or-white fallacy because the premise offers only two choices when in reality there are more.

5. Post Hoc
Premise: Immigration increased, and the economy improved.
Conclusion: Immigration improved the economy.
This is a post hoc fallacy; it reasons that because one event occurred before the other, it caused the second. Correlation does not always indicate causation.

7. Hasty generalization
Premise: My cab driver was rude.
Conclusion: Everyone in this country must be rude.
This is a hasty generalization because it assumes that if something is true for one, then it is true for all.

9. Slippery slope
Premise: Without college you won't get a good job.
Conclusion: You will be on the streets begging for change.
This is a slippery slope because it assumes many factors in the premise are directly connected when the connection between each factor is actually weak. As the connection between factors weakens, so does the argument.

11. Appeal to emotion
Premise: He lost his wife a year ago.
Conclusion: Therefore, if he asks you on a date, you have to say yes.
This is an appeal to emotion because the argument attempts to connect an emotion, pity, to its conclusion.

13. Appeal to popularity
Premise: More than 100,000 people believe that astrology affects people's lives.
Conclusion: There must be some truth to it.
This is an appeal to popularity because the arguer is attempting to connect a popular idea to the reason for the conclusion.

15. Straw man
Premise: Going out of your way is an inconvenience.
Conclusion: There are more important things than convenience.
This is a straw man fallacy because the premise is an attempt to distract the audience from the original statement.

17. Red herring
Premise: Teachers don't get paid enough.
Conclusion: Teachers need to be paid extra for new programs.
This is a red herring fallacy because the premise is an attempt to distract the audience from the original statement.

19. Loaded questions
No premise or conclusion
This is a loaded question because a "yes" or a "no" answer both imply a false or deniable statement.

21. Circular reasoning
Premise: You have human parents.
Conclusion: You are a human being.
This is a circular reasoning because the conclusion is the same as the premise, just reworded.

23. Appeal to common practice
Premise: Everyone drives 75 mph in a 65-mph zone.
Conclusion: Driving 75 mph in a 65-mph zone is acceptable.
This is an appeal to common practice because it relies on general acceptance in its premise rather than logic.

25. Appeal to authority
Premise: The actor stated that tight monetary policy is the best way to shorten a recession.
Conclusion: Tight monetary policy is the best way to shorten a recession.
This is an appeal to authority because it is drawing on expertise from an irrelevant authority.

27. Ad hominem
Premise: You are not a mother.
Conclusion: You don't know what's best for children.
This is an ad hominem fallacy because the premise is an attack on someone's character rather than the issue.

29. Slippery slope
Premise: I should make an exception for you.
Conclusion: I will have to make an exception for everyone.
This is a slippery slope because it assumes many factors in the premise are directly connected, when actually, the connection between each factor is actually weak. As the connection between factors weakens, so does the argument.

31. Appeal to ignorance
Premise: The police could not prove that the strange activity was not from a ghost.
Conclusion: The strange activity was from a ghost.
This is an appeal to ignorance because it is drawing a conclusion based on lack of evidence for the converse of the premise.

33. Red herring
Premise: Students need a greater say in curriculum changes.
Conclusion: Teachers need a greater voice overall.
This is a red herring fallacy because the premise is an attempt to distract the audience from the original statement.

35. Black or white
Premise: Either you love America or you leave.
Conclusion: You must love America so that you don't have to leave.
This is a black-or-white fallacy because the premise offers only two choices when in reality there are more.

37. Appeal to emotion
Premise: My sister is beautiful and intelligent.
Conclusion: Therefore, she should give me a ride to school.
This is an appeal to emotion because the argument attempts to connect an emotion, flattery, to its conclusion.

39. Appeal to common practice
Premise: Most universities compensate athletes.
Conclusion: Compensating athletes is acceptable.
This is an appeal to common practice because it relies on general acceptance in its premise rather than logic.

41. Appeal to popularity
Premise: All of my friends have trucks, not cars.
Conclusion: Trucks are better than cars.
This is an appeal to popularity because the arguer is attempting to connect a popular opinion to the reason for the conclusion.

CHAPTER 2 REVIEW EXERCISES

1. The negative integers greater than –6 are –5, –4, –3, –2, –1. Using the roster method, write the set as {–5, –4, –3, –2, –1}.

2. The set of composite numbers less than 17 is {4, 6, 8, 9, 10, 12, 14, 15, 16}.

3. The set of letters in the name Tennessee is {T, e, n, s}.

4. {April, June, September, November}

5. Adding 4 to each side of the equation produces $x = 7$. {7} is the solution set.

6. Solving:
$$x - 1 < 4$$
$$x < 5$$
The set of whole numbers less than 5 is {0, 1, 2, 3, 4}.

7. $\{2, 4, 6, 7\}' = \{0, 1, 3, 5, 8\}$

8. $\{3, 6\}' = \{0, 1, 2, 4, 5, 7, 8\}$

9. $\varnothing' = U = \{0, 1, 2, 3, 4, 5, 6, 7, 8\}$

10. $\{0, 1, 2, 3, 4, 5, 6, 7, 8\}' = U' = \varnothing$

11. True; every element of F is an element of D.

12. False; r and s are not elements of F.

13. True; s is an element of E and $G \neq E$.

14. True; p and t are elements of D and $F \neq D$.

15. $G' = \{p, q, r, t\}$. Since q is not an element of D, the statement is false.

16. $F' = \{q, r, s\} \neq E$ so the statement is false.

17. False; $\varnothing = \varnothing$ so cannot be a proper subset of itself.

18. False; E does not contain sets, only letters.

19. False; D has 4 elements, $2^4 = 16$ subsets and $2^4 - 1 = 15$ proper subsets.

20. False; $\{0\}$ has 1 element but $\varnothing$ has no elements.

21. $A \cup B = \{2, 4, 6\} \cup \{1, 2, 5, 8\}$
$= \{1, 2, 4, 5, 6, 8\}$

22. $A \cap B = \{2, 4, 6\} \cap \{1, 2, 5, 8\} = \{2\}$

23. $A \cap B' = \{2, 4, 6\} \cap \{3, 4, 6, 7\} = \{4, 6\}$

24. $B \cap C' = \{1, 2, 5, 8\} \cap \{2, 4, 5, 6, 8\}$
$= \{2, 5, 8\}$

25. $(A \cup B)' = (\{1, 2, 4, 5, 6, 8\})' = \{3, 7\}$

26. $A' = \{1, 3, 5, 7, 8\}$.
$A' \cap B = \{1, 3, 5, 7, 8,\} \cap \{1, 2, 5, 8\}$
$= \{1, 5, 8\}$.
$(A' \cap B)' = (\{1, 5, 8\})' = \{2, 3, 4, 6, 7\}$

27. $B \cap C = \{1, 2, 5, 8\} \cap \{1, 3, 7\} = \{1\}$.
$A \cup (B \cap C) = \{2, 4, 6\} \cup \{1\}$
$= \{1, 2, 4, 6\}$

28. $A \cap (B \cup C) = \{2, 4, 6\} \cap \{1, 2, 3, 5, 7, 8\}$
$= \{2\}$

29. $A \cup A' = \{2, 4, 6\} \cup \{1, 3, 5, 7, 8\}$
$= \{1, 2, 3, 4, 5, 6, 7, 8\} = U$

30. $A \cap A' = \{2, 4, 6\} \cap \{1, 3, 5, 7, 8\} = \varnothing$

31. $A' \cap (B \cup C)$

U

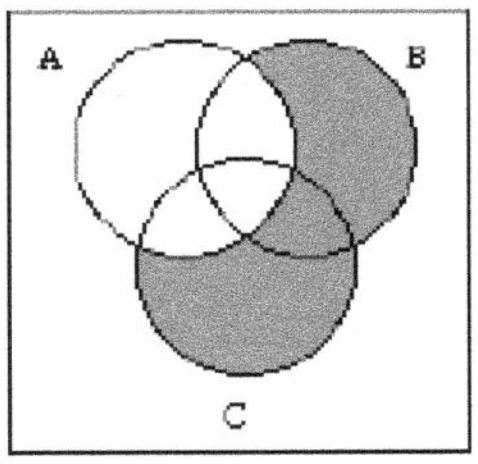

32. $A' \cap B'$

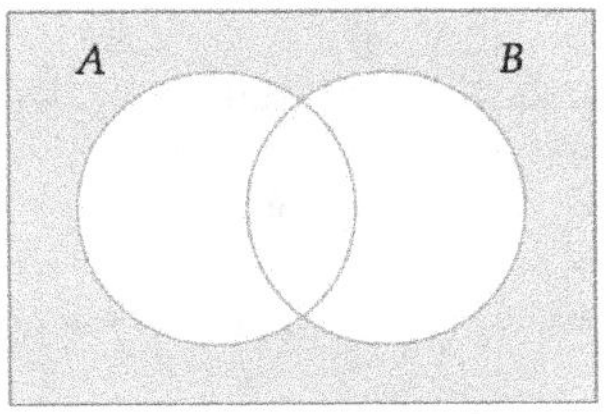

$(A \cap B)'$

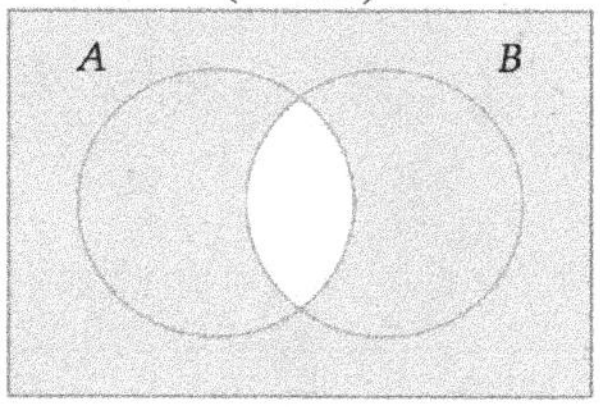

Therefore, they are not equal.

33. There are 5 different ingredients.
$2^5 = 32$ different smoothies

34. $2^9 = 512$ different versions of this sedan

35. Let $M = \{\text{students taking math}\}$ and let $B = \{\text{students who are business majors}\}$
$n(M \cup B) = n(M) + n(B) - n(M \cap B)$
$= 841 + 525 - 202$
$= 1164$
Therefore, 1164 students are taking math or are business majors.

36. Use a Venn diagram to represent the survey results.

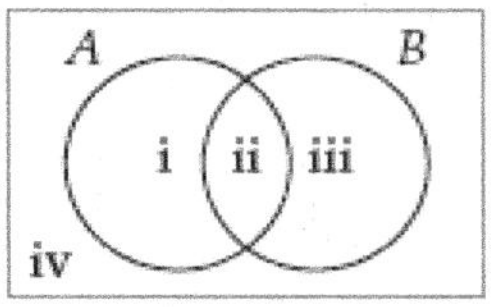

ii: both aerobics and weight training: 97
i: aerobics only: 208 – 97 = 111
iii: weight training only: 145 – 97 = 48
iv: neither: 135
Total the numbers from each region to find the number of members surveyed:
111 + 97 + 48 + 135 = 391

37. a. 148 + 97 + 225 = 470 people
b. 20 people
c. 62 + 78 + 50 = 190 people
d. 148 +62 +78 +20 +97 +50 +225 = 680
750 – 680 = 70 people

38. a. 342 people
b. 78 people
c. 126 + 257 + 198 = 581 people
d. 342+126+257+78+256+198+398=1655
2000 – 1655 = 345 people

39. a. 48+64+58+65+73 = 308 people
b. 72+64+54+56+73 = 319 people
c. 64+72+58+64+65+56+73 = 452 people

d. 65+73 = 138 people

e. 64+(64+72+58+54+56+73)=441 people

40. a.

	Positive Test	Negative Test
Has Mumps Antigen	179	21
Does Not Have Mumps Antigen	83	717

b. 21

c. A person does not have the mumps antigen but tested positive for it.

d. 717

41. a. Some dogs are not poodles.

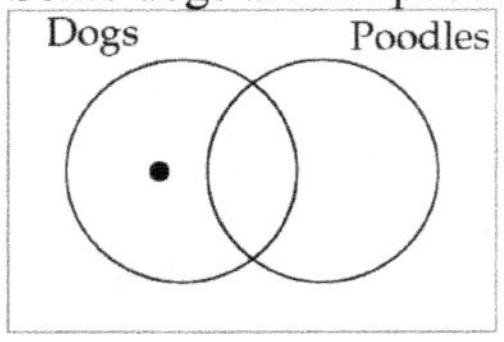

b. All cats are not dogs.

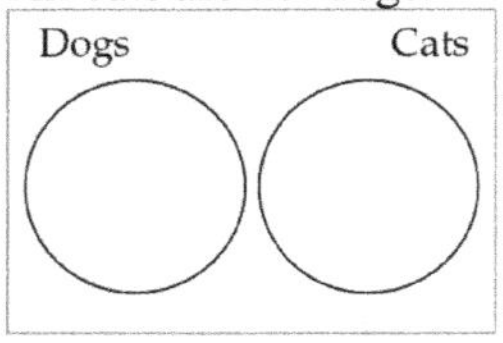

42. a. All cars are fuel efficient.

b. Some parakeets are not birds.

43. The argument is valid.

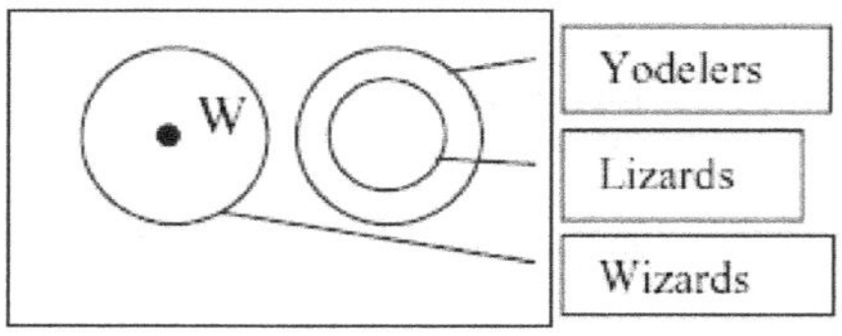

44. The argument is invalid.

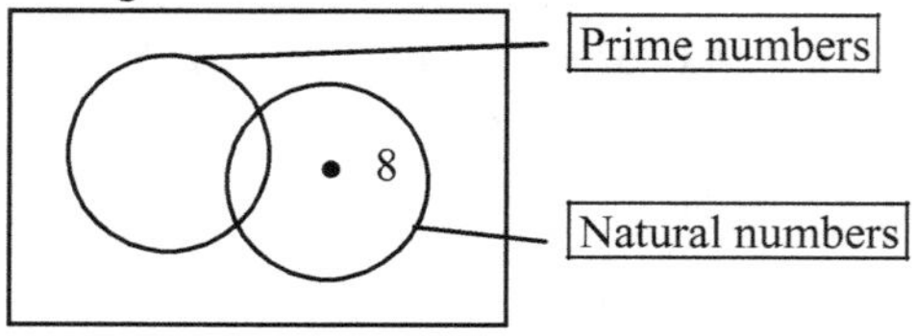

45. The argument is invalid.

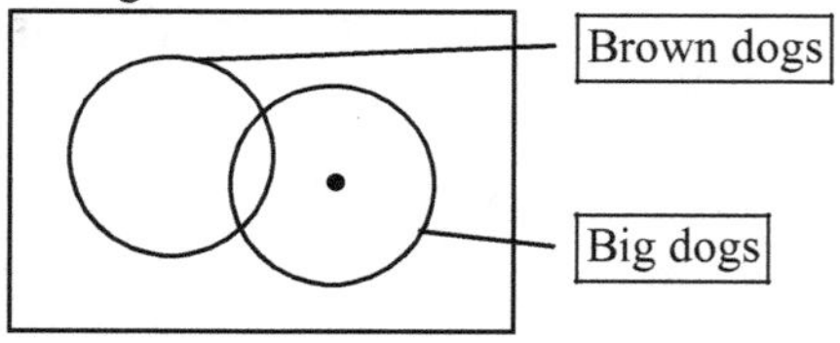

46. The argument is valid.

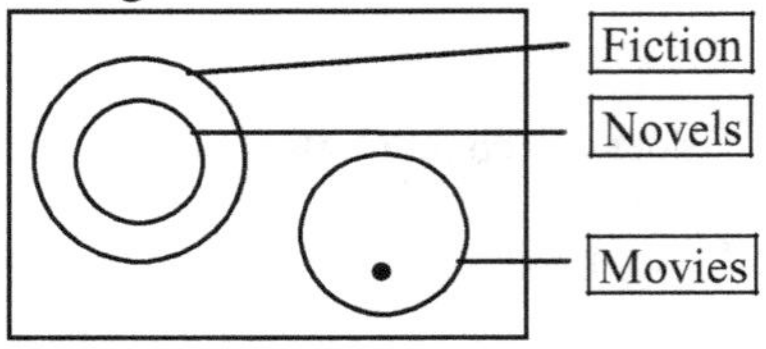

47. The argument is invalid.

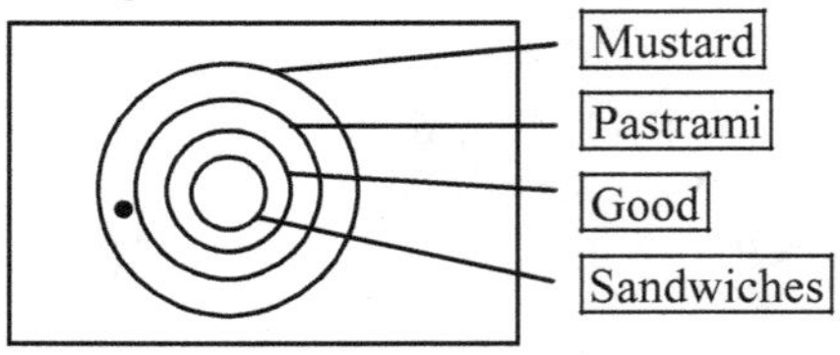

48. The argument is invalid.

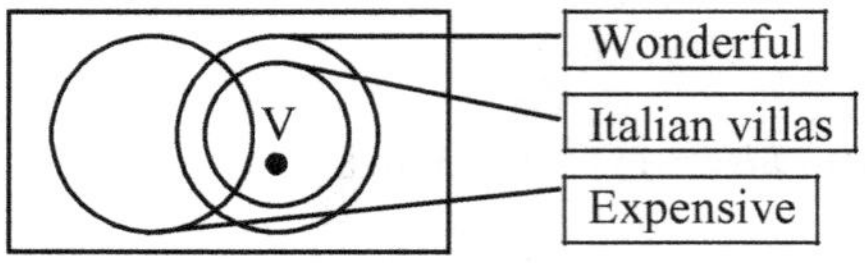

49. The argument is valid.

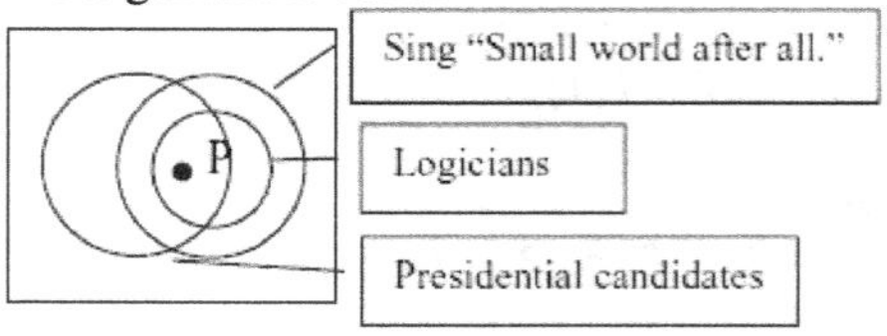

50. Proposition

51. Not a proposition

52. Not a proposition

53. The concert music was not too loud.

54. $p \wedge q$

55. $r \wedge \sim q$

56. $q \rightarrow \sim p$

57. $p \vee q \vee r$

58. The football game will be played on Friday and the football game will be played at home.

59. If the football game is played on Friday, then the visiting team is not from Boston.

60. The football game will be played at home, or the visiting team is from Boston, and the football game will be played on Friday.

61. This is not the time for action, and do not hold your peace.

62. I am not committed to this fine state or will seek the presidential nomination.

63. a. *Antecedent:* I practice my accent.
Consequent: I will sound more like a native speaker.

b. *Antecedent:* Secure a loan on this house.
Consequent: You make a 20% down payment.

64. True. A false antecedent yields a true statement whether the consequent is true or false.

65. False. The antecedent is true and the consequent is false.

66. Both components are always true, so this statement is true.

67. True. A false antecedent yields a true statement whether the consequent is true or false.

68. Converse: If I can ride in the HOV lane, then I have an electric car.
Inverse: If I do not have an electric car, then I cannot ride in the HOV lane.
Contrapositive: If I cannot ride in the HOV lane, then I do not have an electric car.

69. Converse: If I lose some weight, then I start exercising.
Inverse: If I do not start exercising, then I will not lose some weight.
Contrapositive: If I do not lose some weight, then I did not start exercising.

70. Label the statements
f: The shoe fits.
w: Wear the shoe.
In symbolic form the argument is:
In symbolic form:
$f \to w$
$\underline{w\quad}$
$\therefore f$
The argument is invalid using the fallacy of the converse.

71. Label the statements
c: I cut my afternoon class.
p: I will go to the party.
In symbolic form:
$c \to p$
$\underline{\sim p\quad}$
$\therefore \sim c$
The argument is valid using modus tollens.

72. Label the statements
t: I don't watch television.
s: I will fall asleep
w: I will wake up at 3 A.M.
In symbolic form:
$t \to s$
$\underline{s \to w\quad}$
$\therefore \sim w \to \sim t$
The contrapositive of $\sim w \to \sim t$ is $t \to w$
The argument is valid using transitive reasoning.

73. Label the statements
w: I win the lottery.
b: I will buy a Maserati.
In symbolic form:
$w \to b$
$\underline{w\quad}$
$\therefore b$
The argument is valid using modus ponens.

74. This is an ad hominem (personal attack) fallacy.

75. This is circular reasoning.

76. This is a slippery slope fallacy.

77. This is an appeal to emotion fallacy.

Chapter 3: Proportions and Variation

Think About It, Section 3.1

1. 25 blue jelly beans

2. Yes

3. 67

4. $22 per thousand

EXERCISE SET 3.1

1. Examples may vary. For instance, miles per gallon and cost per pound.

3. $\dfrac{1600}{4800} = \dfrac{1}{3}$

5. $\dfrac{300}{800} = \dfrac{3}{8}$

7. $\dfrac{220{,}000 - 180{,}000}{180{,}000} = \dfrac{2}{9}$

9. $\dfrac{582 \text{ miles}}{12 \text{ hours}} = \dfrac{48.5 \text{ miles}}{1 \text{ hour}} = 48.5 \text{ mph}$

11. $\dfrac{544 \text{ words}}{8 \text{ minutes}} = \dfrac{68 \text{ words}}{1 \text{ minute}} = 68 \text{ words/min}$

13. $\dfrac{\$9100}{350 \text{ shares}} = \dfrac{\$26}{1 \text{ share}} = \$26 \text{ per share}$

14. $\dfrac{1000 \text{ square feet}}{2.5 \text{ gallons}} = \dfrac{400 \text{ square feet}}{1 \text{ gallon}} = 400$ square feet per gallon

15. $\dfrac{\$682.50}{35 \text{ hours}} = \dfrac{\$19.50}{1 \text{ hour}} = \$19.50\text{/hr}$

17. a. $5.6 \text{ feet/second}\left(\dfrac{60 \text{ seconds}}{1 \text{ minute}}\right) = 336 \text{ ft/min}$

    b. Using ratios:
    $$\frac{5.6 \text{ feet}}{1 \text{ second}} = \frac{500 \text{ feet}}{x \text{ seconds}}$$
    $$5.6x = 500$$
    $$x \approx 89$$
    89 seconds

19. $\dfrac{\$3.49}{18 \text{ ounces}} \approx \$0.194 \text{ per ounce}$
    $\dfrac{\$3.89}{24 \text{ ounces}} \approx \$0.162 \text{ per ounce}$
    The 24-ounce box is more economical.

21. a. Babe Ruth $\dfrac{540}{60} = 9$
    9 to 1

    b. Explanations may vary. A lower ratio of at bats to home runs means the batter is more likely to hit a home run.

23. a. $2017: \dfrac{219{,}800{,}000{,}000}{2{,}850{,}000{,}000} \approx 77.1$
    $2019: \dfrac{243{,}900{,}000{,}000}{3{,}192{,}000{,}000} \approx 76.4$
    $2021: \dfrac{275{,}700{,}000{,}000}{3{,}480{,}000{,}000} \approx 79.2$

    b. $\dfrac{79.2}{77.1} \approx 1.03$ times greater

25. a. $\dfrac{\text{Number of students}}{\text{Number of faculty}} = \dfrac{10{,}711}{893} \approx 12$
    12:1, 12 to 1. There are 12 students per faculty member at Villanova University.

    b. Calculate each ratio:
    Swathmore: $\dfrac{1581}{200} \approx 8$
    Rice: $\dfrac{6719}{1049} \approx 6$
    Vassar: $\dfrac{2435}{312} \approx 8$
    UCLA: $\dfrac{41{,}908}{932} \approx 45$
    Villanova: $\dfrac{10{,}711}{893} \approx 12$
    Rice University has the lowest.

27. $\dfrac{16{,}369{,}000}{5{,}307{,}000 + 16{,}369{,}000} \times 1000 \approx 755$
    755 employed women per 1000 women in the workforce.

29. $\dfrac{196{,}000}{325{,}377{,}000} \times 100{,}000 \approx 60$
    60 dentists per 100,000 people.

31. $\dfrac{\$5.06}{\$1000} \times \$313{,}900 \approx \$1588.33$

33. Using ratios:
    $$\frac{18.0774}{\$1} = \frac{5000 \text{ pesos}}{x \text{ dollars}}$$
    $$x = \$276.59$$
    $276.59

35. Using ratios:
    $$\frac{3.54239}{\$1} = \frac{21{,}500{,}000 \text{ shekels}}{x \text{ dollars}}$$
    $$x = \$6{,}069{,}348.66$$
    $6,069,348.66, or about $6.07 million

37. Convert from Israeli shekels to U.S. dollars:

$$\frac{3.54239}{\$1}=\frac{1\text{ shekel}}{x\text{ dollars}}$$
$$x\approx \$0.282295$$

Convert from U.S. dollars to Swedish krona:

$$\frac{8.42484}{\$1}=\frac{x\text{ Swedish krona}}{\$0.282295}$$
$$x\approx 2.378\text{ Swedish krona}$$

1 shekel is equivalent to 2.378 Swedish krona

THINK ABOUT IT, SECTION 3.2

1. $\frac{1\text{ foot}}{12\text{ inches}}$

2. kilogram

3. To convert from millimeters to meters, move the decimal point **3** places to the **left**.

4. 1 meter is longer than 1 yard.

5. 1 liter is larger than 1 quart.

EXERCISE SET 3.2

1. $$6\text{ ft}=6\not{\text{ft}}\times\frac{12\text{ in.}}{1\not{\text{ft}}}=6\times 12\text{ in.}=72\text{ in.}$$

3. $$5\text{ yd}=5\not{\text{yd}}\times\frac{3\not{\text{ft}}}{1\not{\text{yd}}}\times\frac{12\text{ in.}}{1\not{\text{ft}}}=5\times 3\times 12\text{ in.}=180\text{ in.}$$

5. $$64\text{ oz}=64\not{\text{oz}}\times\frac{1\text{ lb}}{16\not{\text{oz}}}=\frac{64\text{ lb}}{16}=4\text{ lb}$$

7. $$1\frac{1}{2}\text{ lb}=\frac{3}{2}\not{\text{lb}}\times\frac{16\text{ oz}}{1\not{\text{lb}}}=\frac{48\text{ oz}}{2}=24\text{ oz}$$

9. $$2\frac{1}{2}\text{ c}=\frac{5}{2}\not{\text{c}}\times\frac{8\text{ fl oz}}{1\not{\text{c}}}=\frac{40\text{ ft oz}}{2}=20\text{ fl oz}$$

11. $$7\text{ gal}=7\not{\text{gal}}\times\frac{4\not{\text{qt}}}{1\not{\text{gal}}}\times\frac{2\text{ pt}}{1\not{\text{qt}}}=7\times 4\times 2\text{ pt}=56\text{ pt}$$

13. $62\text{ cm}=620\text{ mm}$

15. $3.21\text{ m}=321\text{ cm}$

17. $7421\text{ g}=7.421\text{ kg}$

19. $0.45\text{ g}=4.5\text{ dg}$

21. $7.5\text{ ml}=0.0075\text{ L}$

23. $0.435\text{ L}=435\text{ ml}=435\text{ cm}^3$

25. $$145\text{ lb}\approx 145\not{\text{lb}}\times\frac{1\text{ kg}}{2.2\not{\text{lb}}}=\frac{145\text{ kg}}{2.2}\approx 65.91\text{ kg}$$
$145\text{ lb}\approx 65.91\text{ kg}$

27. $$\frac{30\text{ mi}}{\text{h}}\approx\frac{30\not{\text{mi}}}{\text{h}}\times\frac{1.61\text{ km}}{1\not{\text{mi}}}=\frac{30\times 1.61\text{ km}}{\text{h}}=48.3\text{ km/h}$$
$30\text{ mi/h}\approx 48.3\text{ km/h}$

29. $$\frac{\$32.99}{\text{gal}}\approx\frac{\$32.99}{\not{\text{gal}}}\times\frac{1\not{\text{gal}}}{3.79\text{ L}}=\frac{\$32.99}{3.79\text{ L}}\approx\frac{\$8.704485}{\text{L}}\approx\frac{\$8.70}{\text{L}}$$
$\$32.99/\text{gal}\approx \$8.70/\text{L}$

31. $$35\text{ mm}=3.5\text{ cm}=3.5\not{\text{cm}}\times\frac{1\text{ in.}}{2.54\not{\text{cm}}}=\frac{3.5\text{ in.}}{2.54}\approx 1.37795\text{ in.}\approx 1.38\text{ in.}$$
$35\text{ mm}\approx 1.38\text{ in.}$

33. $$\frac{\$10}{\text{kg}}\approx\frac{\$10}{\not{\text{kg}}}\times\frac{1\not{\text{kg}}}{2.2\text{ lb}}=\frac{\$10}{2.2\text{ lb}}\approx\frac{\$4.545455}{\text{lb}}\approx\frac{\$4.55}{\text{lb}}$$
$\$10/\text{kg}\approx \$4.55/\text{lb}$

THINK ABOUT IT, SECTION 3.3

1. 3 and b are the extremes; 4 and a are the means.

2. For a proportion, the product of the means **equals** the product of the extremes.

3. Yes

4. No. For instance, it is possible that $a = 6$ and $b = 8$.

5. The corresponding units do not match.

EXERCISE SET 3.3

1. Solving:
$$\frac{3}{8} = \frac{x}{12}$$
$$8x = 36$$
$$x = 4.5$$

3. Solving:
$$\frac{7}{12} = \frac{25}{d}$$
$$7d = 300$$
$$d \approx 42.86$$

5. Solving:
$$\frac{15}{45} = \frac{72}{c}$$
$$15c = 3240$$
$$c = 216$$

7. Solving:
$$\frac{65}{20} = \frac{14}{a}$$
$$65a = 280$$
$$a \approx 4.31$$

9. Solving:
$$\frac{0.5}{2.3} = \frac{b}{20}$$
$$2.3b = 10$$
$$b \approx 4.35$$

11. Solving:
$$\frac{0.7}{1.2} = \frac{6.4}{x}$$
$$0.7x = 7.68$$
$$x \approx 10.97$$

13. Solving:
$$\frac{1}{6} = \frac{x}{174}$$
$$6x = 174$$
$$x = 29$$
29 pounds

15. Solving:
$$\frac{2}{80} = \frac{x}{220}$$
$$80x = 440$$
$$x = 5.5$$
5.5 mg

17. Solving:
$$\frac{1}{3} = \frac{5}{x}$$
$$x = 15$$
$$\frac{1}{3} = \frac{8}{y}$$
$$y = 24$$
15 feet by 24 feet

19. Solving:
$$\frac{3}{4} = \frac{48}{x}$$
$$3x = 192$$
$$x = 64$$
64 inches

21. Solving:
$$\frac{2}{3} = \frac{x}{240{,}000}$$
$$3x = 480{,}000$$
$$x = 160{,}000$$
160,000 people

23. The amount of fat in a pancake is proportional to the area of the pancake, which is proportional to the square of the pancake's diameter.
$$\frac{4^2}{5} = \frac{6^2}{x}$$
$$16x = 180$$
$$x = 11.25$$
11.25 grams

25. $32 + 8 + 1 = 41$
Total weight = 50 pounds
Let x be the number of pounds of flour.
$$\frac{32}{41} = \frac{x}{50}$$
$$1600 = 41x$$
$$39.0 \approx x$$
Let y be the number of pounds of sugar.
$$\frac{8}{41} = \frac{y}{50}$$
$$400 = 41y$$
$$9.8 \approx y$$
Let z be the number of pounds of water.
$$\frac{1}{41} = \frac{z}{50}$$
$$50 = 41z$$
$$1.2 \approx z$$
The chef needs about 39.0 pounds of flour, 9.8 pounds of sugar, and 1.2 pounds of water.

THINK ABOUT IT, SECTION 3.4

1. Greater than. $0.2x = 16$, so $x = 80 > 16$.
2. Less than. $1.50x = 90$, so $x = 60 < 90$.
3. Less than.
4. There may be people who favor both.
5. The five scores of 90% count more heavily than one score of 86%.

EXERCISE SET 3.4

1. $40\% = 0.40$
   $40\% = 40\left(\frac{1}{100}\right) = \frac{40}{100} = \frac{2}{5}$

3. $73\% = 0.73$
   $73\% = 73\left(\frac{1}{100}\right) = \frac{73}{100}$

5. $2\% = 0.02$
   $2\% = 2\left(\frac{1}{100}\right) = \frac{2}{100} = \frac{1}{50}$

7. $0.5\% = 0.005$
   $0.5\% = 0.5\left(\frac{1}{100}\right) = \frac{0.5}{100} = \frac{1}{200}$

9. $1.5\% = 0.015$
   $1.5\% = 1.5\left(\frac{1}{100}\right) = \frac{1.5}{100} = \frac{3}{200}$

11. $56.25\% = 0.5625$
    $56.25\% = 56.25\left(\frac{1}{100}\right) = \frac{56.25}{100} = \frac{9}{16}$

13. $\frac{3}{5} = 0.6 = 60\%$

15. $\frac{3}{8} = 0.375 = 37.5\%$

17. $0.35 = 35\%$

19. $0.6 = 60\%$

21. $0.05 = 5\%$

23. $\frac{3}{200} = 0.015 = 1.5\%$

25. 92% is 0.92 Solving:
    $(0.92)(136) = A$
    $A = 125.12 \approx 125$
    125 million returns were filed electronically.

27. Solving:
    $(0.15)(390) = A$
    $58.5 = A$
    \$58.5 billion

29. Solving:
    $\frac{\text{Percent}}{100} = \frac{\text{amount}}{\text{base}}$
    $\frac{p}{100} = \frac{293}{1236}$
    $1236p = 29{,}300$
    $p \approx 23.7$
    23.7% were irked most by tailgaters.

31. Solving:
    $\frac{\text{Percent}}{100} = \frac{\text{amount}}{\text{base}}$
    $\frac{16}{100} = \frac{2{,}240{,}000}{B}$
    $16B = 224{,}000{,}000$
    $B = 14{,}000{,}000$
    14,000,000 ounces

33. $\frac{\text{Percent}}{100} = \frac{\text{amount}}{\text{base}}$
    $\frac{22}{100} = \frac{740}{B}$
    $22B = 74{,}000$
    $B \approx 3364$
    3364 people

35. Solving:
    $\frac{p}{100} = \frac{45}{78}$
    $78p = 4500$
    $p \approx 57.7$
    57.7%

37. a. Using ratios:
    $\frac{x}{100} = \frac{8.1 - 4.0}{4.0}$
    $4.0x = 410$
    $x = 102.5$
    102.5% increase

    b. Using ratios:
    $\frac{x}{100} = \frac{18.0 - 8.1}{8.1}$
    $8.1x = 990$
    $x \approx 122.2$
    122.2% increase

    c. Using ratios:
    $\frac{x}{100} = \frac{18.0 - 4.0}{4.0}$
    $4.0x = 1400$
    $x = 350$
    350% increase

d. $\frac{18}{4} = 4.5$ times as large. Convert the percent to a decimal and add 1.

39. Solving:
$$0.30(22.2) = A$$
$$6.66 = A$$
$$22.2 + 6.66 = 28.86$$
28.86 million

41. a. Using ratios:
$$\frac{x}{100} = \frac{9,980,000 - 9,620,000}{9,620,000}$$
$$9,620,000x = 36,000,000$$
$$x \approx 3.7$$
3.7% increase

b. Solving:
$$0.037(9,980,000) = A$$
$$369,260 = A$$
$$9,980,000 + 369,260 = 10,349,260$$
$$\approx 10,349,000$$
10,349,000 millionaires

43. $(1.05)(1.06)(1.07) < (1.06)^3$ Your salary is less.

THINK ABOUT IT, SECTION 3.5

1. If y varies directly as x, then as x increases, y **increases**.

2. $y = kx^2$

3. If y varies inversely as x, then as x decreases, y **increases**.

4. $y = \frac{k}{x^2}$

5. Directly

6. Inversely

EXERCISE SET 3.5

1. Write the equation and solve for k.
$$y = kx$$
$$64 = k \cdot 48$$
$$\frac{64}{48} = k$$
$$\frac{4}{3} = k$$

3. Write the equation and solve for k.
$$r = kt^2$$
$$144 = k \cdot 108^2$$
$$\frac{144}{108^2} = k$$
$$\frac{2^4 \cdot 3^2}{2^4 \cdot 3^6} = k$$
$$\frac{1}{81} = k$$

5. Write the equation and solve for k.
$$T = krs^2$$
$$210 = k \cdot 30 \cdot 5^2$$
$$\frac{210}{30 \cdot 5^2} = k$$
$$\frac{7}{25} = k$$
$$0.28 = k$$

7. Write the equation and solve for k.
$$V = klwh$$
$$240 = k \cdot 8 \cdot 6 \cdot 5$$
$$\frac{240}{8 \cdot 6 \cdot 5} = k$$
$$1 = k$$

9. Find the volume of the balloon.
$$V = kT$$
$$0.85 = k \cdot 270$$
$$\frac{0.85}{270} = k$$
$$\frac{0.17}{54} = k$$
Thus $V = \frac{0.17}{54}T = \frac{0.17}{54} \cdot 324 = (0.17)6$
$$= 1.02 \text{ liters}$$

11. Find the number of semester hours.
$$s = k \cdot q$$
$$34 = k \cdot 51$$
$$\frac{2}{3} = k$$
$$p = \frac{2}{3} \cdot 93$$
$p = 62$ semester hours

13. Find the amount of juice. Round to the nearest tenth of a fluid ounce.

$$j = k \cdot d^3$$
$$6 = k \cdot (4)^3$$
$$\frac{3}{32} = k$$
$$p = \frac{3}{32} \cdot (5)^3$$
$$p \approx 11.7 \text{ fl oz}$$

15. $T = k\sqrt{l}$

$$1.8 = k\sqrt{3}$$
$$k = \frac{1.8}{\sqrt{3}} \approx 1.03923$$

a. $T = \frac{1.8}{\sqrt{3}}\sqrt{10} = \frac{1.8\sqrt{30}}{3} = 0.6\sqrt{30}$

≈ 3.3 seconds

b.
$$T = k\sqrt{l}$$
$$\frac{T}{k} = \sqrt{l}$$
$$\frac{2}{1.03923} = \sqrt{l}$$
$$l = \frac{4}{1.03923^2} \approx 3.7 \text{ ft}$$

17. Find the speed of the gear with 48 teeth.

$$r = \frac{k}{t}$$
$$30 = \frac{k}{64}$$
$$1920 = k$$
$$r = \frac{1920}{48}$$
$r = 40$ revolutions per minute

19. Find the sound intensity.

$$I = \frac{k}{d^2}$$
$$0.5 = \frac{k}{7^2}$$
$$24.5 = k$$
$$I = \frac{24.5}{d^2}$$
$$I = \frac{24.5}{10^2}$$
$$I = 0.245 \text{ W/m}^2$$

21. a.
$$V = kr^2h$$
$$V_1 = k(3r)^2h$$
$$= 9(kr^2h)$$
$$= 9V$$
Thus the new volume is 9 times the original volume.

b.
$$V_2 = kr^2(3h)$$
$$= 3(kr^2h)$$
$$= 3V$$
Thus the new volume is 3 times the original volume.

c.
$$V_3 = k(3r)^2(3h)$$
$$= k9r^2 \cdot 3 \cdot h$$
$$= 27(kr^2h)$$
$$= 27V$$
Thus the new volume is 27 times the original volume.

23. Find what happens to the volume.

$$V = \frac{knT}{P}$$
$$V_1 = \frac{k(3n)T}{\left(\frac{1}{2}p\right)}$$
$$= 6\left(\frac{knT}{p}\right)$$
$$= 6V$$
Thus the new volume is 6 times the original volume.

25. Find the force. Round to the nearest 10 lb.

$$F = \frac{kws^2}{r}$$
$$2800 = \frac{k \cdot 1800 \cdot 45^2}{425}$$
$$\frac{2800 \cdot 425}{1800 \cdot 45^2} = k$$
$$\frac{14 \cdot 425}{9 \cdot 45^2} = k$$
$$0.3264746 \approx k$$
Thus $F = \frac{(0.3264746) \cdot 1800 \cdot 55^2}{450}$

≈ 3950 pounds

27. Find the depth.

$$S = kwd^3$$
when $d = 10,\ w = \sqrt{18^2 - 10^2} \approx 15$

$$S \approx k(15)(10)^3 = 15{,}000k$$
$$d = 12,\ w = \sqrt{18^2 - 12^2} \approx 13.4$$
$$S \approx k(13.4)(12)^3 = 23{,}155k$$
$$d = 14,\ w = \sqrt{18^2 - 14^2} \approx 11.3$$
$$S \approx k(11.3)(14)^3 = 31{,}007k$$
$$d = 16,\ w = \sqrt{18^2 - 16^2} \approx 8.2$$
$$S \approx k(8.2)(16)^3 = 33{,}587k$$
The strongest beam occurs when $d = 16$ inches.

CHAPTER 3 REVIEW EXERCISES

1. $\frac{350,000-280,000}{280,000}=\frac{70,000}{280,000}=\frac{1}{4}$

2. $\frac{6 \text{ feet}}{10 \text{ inches}}=\frac{72 \text{ inches}}{10 \text{ inches}}=7.2$

3. $\frac{29,524 \text{ students}}{2405 \text{ teachers}}\approx\frac{12.3 \text{ students}}{1 \text{ teacher}}=12.3 \text{ to } 1$

4. $\frac{\$1896}{\$4395}\approx 0.43$
   Greater than

5. $\frac{326.6 \text{ miles}}{11.5 \text{ gallons}}=\frac{28.4 \text{ miles}}{1 \text{ gallon}}=28.4$ miles per gallon of gasoline

6. $\frac{\$4.23}{20 \text{ ounces}}\approx \$0.212$ per ounce
   $\frac{\$3.75}{18 \text{ ounces}}\approx \$0.208$ per ounce
   The 18-ounce box for $3.75 is more economical.

7. New York: $\frac{8,400,000}{321.8}\approx 26,103$
   Los Angeles: $\frac{3,900,000}{467.4}\approx 8344$
   Chicago: $\frac{2,900,000}{228.469}\approx 12,693$
   Houston: $\frac{2,300,000}{594.03}\approx 3872$
   New York, Chicago, Los Angeles, Houston

8. $\frac{270}{1,200,000}\times 100,000=22.5$
   22.5 pedestrian accidents per 100,000 people.

9. $\frac{27}{100,000}\times 1,271,000\approx 343$
   343 people

10. $\frac{\$22.38}{\$1000}\times \$125,000=\$2797.50$

11. Using ratios:
    $$\frac{64.6108}{\$1}=\frac{16,000,000 \text{ Indian rupees}}{x \text{ dollars}}$$
    $$x\approx \$247,636.62$$
    $247,636.62

12. $27 \text{ in.}=27 \cancel{\text{in.}}\times\frac{1 \text{ ft}}{12 \cancel{\text{in.}}}=2\frac{1}{4} \text{ ft}$

13. $2.5 \text{ lb}=2.5 \cancel{\text{lb}}\times\frac{16 \text{ oz}}{1 \cancel{\text{lb}}}=40 \text{ oz}$

14. $16 \text{ c}=16 \cancel{\text{c}}\times\frac{1 \text{ pt}}{2 \cancel{\text{c}}}\times\frac{1 \text{ qt}}{2 \cancel{\text{pt}}}=4 \text{ qt}$

15. 37 mm = 3.7 cm

16. 3.2 kg = 3,200,000 mg

17. 5.8 dL = 0.58 L

18. $\frac{\$3.56}{\text{L}}\approx\frac{\$3.56}{\cancel{\text{L}}}\times\frac{1 \cancel{\text{L}}}{1.06 \text{ qt}}\approx \$3.36/\text{qt}$

19. $\frac{\$2.45}{\text{gal}}\approx\frac{\$2.45}{1 \cancel{\text{gal}}}\times\frac{1 \cancel{\text{gal}}}{3.79 \text{ L}}\approx \$0.65/\text{L}$

20. $\frac{45 \text{ miles}}{\text{hour}}\approx\frac{45 \cancel{\text{miles}}}{1 \text{ hour}}\times\frac{1.61 \text{ km}}{1 \cancel{\text{mile}}}\approx 72 \text{ km/hr}$

21. Solving:
    $$\frac{n}{4}=\frac{5}{12}$$
    $$12n=20$$
    $$n\approx 1.7$$

22. Solving:
    $$\frac{6}{x}=\frac{3}{7}$$
    $$42=3x$$
    $$14=x$$

23. Solving:
    $$\frac{27}{25}=\frac{n}{9}$$
    $$243=25n$$
    $$n\approx 9.7$$

24. Solving:
    $$\frac{9}{2}=\frac{6}{p}$$
    $$9p=12$$
    $$p\approx 1.3$$

25. Using ratios:
    $$\frac{3}{4}=\frac{t}{10}$$
    $$4t=3(10)=30$$
    $$t=7.5$$
    7.5 tablespoons of fertilizer are required.

26. Using ratios:

$$\frac{150}{2}=\frac{t}{7}$$
$$2t=1050$$
$$t=525\text{ gallons}$$

Find the unit rate.

$$\frac{150}{2}=75\text{ gal/min}$$
$$\frac{3750}{75}=50\text{ min}$$

27. Solving:

$$\frac{28}{200}=\frac{173}{x}$$
$$28x=34,600$$
$$x\approx 1235.7$$

1236 elk

28. Solving:

$$\frac{2}{1+2+3}=\frac{x}{240,000}$$
$$6x=480,000$$
$$x=80,000$$

$80,000

29. $4\%=0.04$

$$4\%=4\left(\frac{1}{100}\right)=\frac{4}{100}=\frac{1}{25}$$

30. $23\%=0.23$

$$23\%=23\left(\frac{1}{100}\right)=\frac{23}{100}$$

31. $12.5\%=0.125$

$$12.5\%=12.5\left(\frac{1}{100}\right)=\frac{12.5}{100}=\frac{1}{8}$$

32. $150\%=1.5$

$$150\%=150\left(\frac{1}{100}\right)=\frac{150}{100}=\frac{3}{2}$$

33. $0.08=8\%$

34. $1.2=120\%$

35. $\frac{3}{200}=0.015=1.5\%$

36. $\frac{5}{8}=0.625=62.5\%$

37. Solving:

$$(0.65)(35,000)=A$$
$$22,750=A$$

22.750 strikes

38. $\frac{\text{Percent}}{100}=\frac{\text{amount}}{\text{base}}$

$$\frac{9}{100}=\frac{18}{B}$$
$$9B=1800$$
$$B=200$$

200 students

39. Solving:

$$\frac{p}{100}=\frac{47.1}{125}$$
$$125p=4710$$
$$p\approx 37.7$$

37.7%

40. Using ratios:

$$\frac{x}{100}=\frac{13-11}{11}$$
$$11x=200$$
$$x=18.2$$

18.2% increase

41. Using ratios:

$$\frac{x}{100}=\frac{95-88}{95}$$
$$95x=700$$
$$x=7.4$$

7.4% decrease

42. Solving:

$$(0.11)(64,500)=A$$
$$7095=A$$

$7095

43. a. $\frac{170,931}{170,931+164,119}\approx 0.51=51.0\%$

51.0% of the projected population in 2025 is female.

b. 2050:

$$\frac{200,696}{200,696+193,234}\approx 0.509=50.9\%$$

Less than one percent.

44. $\frac{27,222-26,400}{27,222}\approx 0.030=3.0\%$

45. Find the acceleration.

$$F=ka$$
$$10=k(2)$$
$$5=k$$
$$F=5a$$
$$15=5a$$
$$3=a$$
$$a=3\text{ ft/s}^2$$

46. Find the distance.

$$d = kt^2$$
$$10.6 = k(2)^2$$
$$2.65 = k$$
$$d = 2.65t^2$$
$$d = 2.65(3)^2$$
$$d = 23.85 \text{ ft}$$

47. Find the number of video games sold if the price is \$30.

$$n = \frac{k}{p}$$
$$5000 = \frac{k}{45}$$
$$225{,}000 = k$$
$$n = \frac{225{,}000}{30}$$
$$n = 7500 \text{ video games sold}$$

48. Find the repulsive force.

$$F = \frac{k}{d^2}$$
$$40 = \frac{k}{(2)^2}$$
$$160 = k$$
$$F = \frac{160}{d^2}$$
$$F = \frac{160}{4^2}$$
$$F = 10 \text{ lb}$$

49. Find the resistance force.

$$F = kSv$$
$$50 = k \cdot 20 \cdot 30$$
$$\frac{50}{20 \cdot 30} = k$$
$$\frac{1}{12} = k$$
$$\text{Thus } F = \frac{1}{12} \cdot 20 \cdot 45 = 75 \text{ pounds}$$

50. Find the volume.

$$V = \frac{kT}{P}$$
$$200 = \frac{k \cdot 100}{20}$$
$$40 = k$$
$$\text{Thus } V = \frac{40 \cdot 120}{30} = 160 \text{ cubic centimeters}$$

Chapter 4: Managing Your Finances

Think About It, Section 4.1

1. Yes

2. Present value, \$500; Maturity value \$525

3. \$20

4. Raul
   Raul will earn \$10 interest. Helena will earn \$5 interest. Raul's interest rate must be higher.

Exercise Set 4.1

1. First find the amount of interest paid. Subtract the principal from the maturity value.
   $I = A - P = 5125 - 5000 = 125$
   Find the simple interest rate by solving the simple interest formula for r.
   $$I = Prt$$
   $$125 = 5000r\left(\frac{3}{12}\right)$$
   $$125 = 1250r$$
   $$0.10 = r$$
   $$r = 10\%$$
   The simple interest rate on the loan is 10%.

3. Using the formula:
   $$A = P(1+rt) = 880[1+0.092(1)]$$
   $$= 880(1.092) = 960.96$$
   The future value of the investment is \$960.96.

5. $A = P(1+rt) = 132\left[1+0.09\left(\frac{1}{12}\right)\right] = 132.99$
   You will owe \$132.99.

7. $A = P(1+rt) = 1200\left[1+0.11\left(\frac{2}{12}\right)\right] = 1222$
   You will owe \$1222.

9. $I = Prt = 1600(0.09)\left(\frac{45}{360}\right) = 18$
   The simple interest earned is \$18.

11. $I = Prt = 4800(0.0725)\left(\frac{150}{360}\right) = 145$
    The simple interest due is \$145.

13. Use 360 days a year.
    $$A = P(1+rt) = 10{,}000\left[1+0.09\left(\frac{140}{360}\right)\right]$$
    $$= 10{,}350$$
    The maturity value is \$10,350.

15. $P = 2000$, $r = 0.085$, and $t = \frac{3}{12}$ year.
    $$A = Prt = 2000\left[0.085\left(\frac{3}{12}\right)\right] = 42.50$$
    The interest penalty is \$42.50.

17. Solve the simple interest formula for r.
    $$I = Prt$$
    $$4(17.50) = 400r\left(\frac{18}{360}\right)$$
    $$70 \cdot \frac{1}{20} = r$$
    $$3.5 = r$$
    $$r = 350\%$$
    The simple interest rate is 350%.

19. Solve the maturity value formula for r.
    $$A = P + Prt$$
    $$5000 = 4900 + 4900r\left(\frac{3}{12}\right)$$
    $$5000 = 4900 + 1225r$$
    $$100 = 1225r$$
    $$\frac{100}{1225} = r$$
    $$r \approx 0.082$$
    The simple interest rate is 8.2%.

21. Solve the maturity value formula for r.
    $$A = P + Prt$$
    $$6000 = 5945 + 5945r\left(\frac{125}{360}\right)$$
    $$6000 = 5945 + \frac{148{,}625}{72}r$$
    $$55 = \frac{148{,}625}{72}r$$
    $$55 \cdot \frac{72}{148{,}625} = r$$
    $$r \approx 0.027$$
    The simple interest rate is 2.7%.

23. $I = Prt = 650(0.033)\left(\frac{35}{360}\right) \approx 2.09$
    The simple interest due is \$2.09.

25. $I = Prt = 725(0.015)(4) = 43.50$
    The simple interest due is \$43.50.

27. Using the formula:
    $$A = P(1+rt) = 2300\left[1+0.0225\left(\frac{8}{12}\right)\right]$$
    $$= 2300(1.015) = 2334.50$$
    The future value of the investment is \$2334.50.

THINK ABOUT IT, SECTION 4.2

1. Compounded daily
2. Present value, \$5000; Future value \$5100
3. Nominal \$100; real \$94
4. Nominal, 5%; Annual percentage yield 5.13%

EXERCISE SET 4.2

1. $P = 5000$, $r = 5.1\% = 0.051$, $n = 12$, $t = 7$

$$A = P\left(1+\frac{r}{n}\right)^{nt}$$
$$= 5000\left(1+\frac{0.051}{12}\right)^{(12\cdot 7)}$$
$$= 5000(1.00425)^{84} \approx 7139.78$$

The compound amount after 7 years is approximately \$7139.78.
Calculate the interest earned.
$I = A - P = 7139.78 - 5000 = 2139.78$
The amount of interest earned is approximately \$2139.78.

3. Interest is compounded daily.
$P = 20{,}000$, $n = 365$, $r = 2.75\% = 0.0275$, $t = 4$

$$A = P\left(1+\frac{r}{n}\right)^{nt}$$
$$= 20{,}000\left(1+\frac{0.0275}{365}\right)^{(365\cdot 4)}$$
$$\approx 22{,}325.47$$

The compound amount after 4 years is approximately \$22,325.47.

5. $A = 10{,}000$, $r = 7\% = 0.07$, $n = 4$, $t = 5$

$$P = \frac{A}{\left(1+\frac{r}{n}\right)^{nt}}$$
$$= \frac{10{,}000}{\left(1+\frac{0.07}{4}\right)^{(4\cdot 5)}}$$
$$= \frac{10{,}000}{(1.0175)^{20}}$$
$$\approx 7068.25$$

\$7068.25 should be invested in the account to have \$10,000 in 5 years.

7. From 2010: $51{,}000\cdot\frac{237.017}{218.056} \approx \$55{,}434.69$

From 2014: $55{,}000\cdot\frac{237.017}{236.736} \approx \$55{,}065.28$

From 2016: $62{,}000\cdot\frac{237.017}{240.008} \approx \$61{,}227.35$

9. Find the future value of \$100 after 1 year.
$P = 100$, $r = 3\% = 0.03$, $n = 12$

$$A = P\left(1+\frac{r}{n}\right)^{nt}$$
$$= 100\left(1+\frac{0.03}{12}\right)^{(12\cdot 1)}$$
$$= 100(1.0025)^{12} \approx 103.04$$

Find the interest earned on the \$100.
$I = A - P = 103.04 - 100 = 3.04$
The effective interest rate is 3.04%.

11. $P = 3115$, $r = 7\% = 0.07$, $n = 1$, $t = 15$

$$A = P\left(1+\frac{r}{n}\right)^{nt}$$
$$= 3115\left(1+\frac{0.07}{1}\right)^{(1\cdot 15)}$$
$$= 3115(1.07)^{15} \approx 8594.38$$

Fifteen years from now, the average rent for an apartment will be approximately \$8594.38.

13. $$\text{Inflation rate} = 100\left(\frac{\text{CPI in 2012} - \text{CPI in 2010}}{\text{CPI in 2010}}\right)$$
$$= 100\left(\frac{229.594 - 218.056}{218.056}\right)$$
$$\approx 5.2913$$

The inflation rate is approximately 5.3%.

15. Calculate the compound amount.
$P = 20{,}000$, $r = 4\% = 0.04$, $n = 12$, $t = 6$

$$A = P\left(1+\frac{r}{n}\right)^{nt}$$
$$= 20{,}000\left(1+\frac{0.04}{12}\right)^{(12\cdot 6)}$$
$$= 20{,}000(1.00\overline{3})^{10} \approx 25{,}414.84$$

Calculate the interest earned.
$I = A - P = 25{,}414.84 - 20{,}000 = 5414.84$
The amount of interest earned is approximately \$5414.84.

17. $P = 4000$, $r = 6\% = 0.06$, $n = 12$, $t = 6$

$$A = P\left(1+\frac{r}{n}\right)^{nt}$$
$$= 4000\left(1+\frac{0.06}{12}\right)^{(12\cdot 6)}$$
$$= 4000(1.005)^{72} \approx 5728.18$$

The future value is approximately \$5728.18.

19. Calculate the compound amount.
$P = 2000$, $r = 6\% = 0.06$, $n = 4$, $t = 3$

$$A = P\left(1+\frac{r}{n}\right)^{nt}$$
$$= 2000\left(1+\frac{0.06}{4}\right)^{(4\cdot 3)}$$
$$= 2000(1.015)^{12} \approx 2391.24$$

Calculate the interest earned.
$I = A - P = 2391.24 - 2000 = 391.24$
The amount of interest earned is approximately \$391.24.

21. $A = 40{,}000$, $r = 8\% = 0.08$, $n = 4$, $t = 18$

$$P = \frac{A}{\left(1+\frac{r}{n}\right)^{nt}} = \frac{40{,}000}{\left(1+\frac{0.08}{4}\right)^{(4\cdot 18)}} = \frac{40{,}000}{(1.02)^{72}} \approx 9612.75$$

\$9612.75 should be invested in the account to have \$40,000 in 18 years.

23. $A = 1{,}000{,}000$, $r = 9\% = 0.09$, $n = 365$, $t = 30$

$$P = \frac{A}{\left(1+\frac{r}{n}\right)^{nt}} = \frac{1{,}000{,}000}{\left(1+\frac{0.09}{365}\right)^{(365\cdot 30)}} \approx 67{,}227.88$$

\$67,227.88 should be invested in the account in order to have \$1,000,000 in 30 years.

25. Calculate the compound amount for the first year. $P = 5000$, $r = 1.5\% = 0.015$, $n = 365$, $t = 1$

$$A = P\left(1+\frac{r}{n}\right)^{nt} = 5000\left(1+\frac{0.015}{365}\right)^{(365\cdot 1)} \approx 5075.56$$

Calculate the compound amount for the next two years. The principal is now \$5075.56 + \$5000 = \$10,075.56.
$r = 1.75\% = 0.0175$, $n = 365$, $t = 1$

$$A = P\left(1+\frac{r}{n}\right)^{nt} = 10{,}075.56\left(1+\frac{0.0175}{365}\right)^{(365\cdot 1)} \approx 10{,}253.43$$

After 2 years, the value is \$10,253.43.

27. $n = 2$, APR $= 4.875\% = 0.04875$

$$\text{APY} = \left(1+\frac{\text{APR}}{n}\right)^{n} - 1 = \left(1+\frac{0.04875}{2}\right)^{2} - 1 \approx 0.0493441406$$

The APY is approximately 4.93%.

THINK ABOUT IT, SECTION 4.3

1. Payment for an ordinary annuity is made at the end of the payment period; payment for a due annuity is made at the beginning of the payment period.

2. Answers may vary. A car loan payment is based on an ordinary annuity. The first payment is due at the end of the month. Funding a retirement account is usually a due annuity. When the account is opened, a payment is made, the beginning of the payment period.

3. The due annuity would have the greater value after 10 years. It has one more interest period than an ordinary annuity.

EXERCISE SET 4.3

1. Using a TI-84 Calculator, enter the values of N = 8(12) = 96, I% = 5, temporarily skip PV, PMT = 2500, FV = 0, P/Y and C/Y = 12. Select END since payments are made at the end of each month. Move the cursor to PV and press ALPHA SOLVE.
The present value of the annuity is \$197,473.60.

3. Down payment = 2250(0.30) = 675
Amount financed = 2250 – 675 = 1575
   a. N = 9, I% = 5.5, PV = 1575, temporarily skip PMT, FV = 0, P/Y and C/Y = 12, END. Move the cursor to PMT and press ALPHA SOLVE.
   The monthly payment is \$179.03.
   b. Total paid = 179.03(9) = 1611.27
   Amount of interest
   = 1611.27 – 1575 = 36.27
   The amount of interest is \$36.27.

5. First find Logan's monthly payment amount. N = 18, I% = 3.5, PV = 1500, temporarily skip PMT, FV = 0, P/Y and C/Y = 12, END.
Move the cursor to PMT and press ALPHA SOLVE.
The monthly payment is \$85.66.

Since Logan has made 12 payments, N = 18 – 12 = 6. Then I% = 3.5, temporarily skip PV, PMT = –85.66, FV = 0, P/Y and C/Y = 12, END.
Move the cursor to PV and press ALPHA SOLVE.
Logan's payoff is \$508.75.

7. N = 20(12) = 240, I% = 6.4, PV = 0, PMT = 250, temporarily skip FV, P/Y and C/Y = 12, BEGIN.
Move the cursor to FV and press ALPHA SOLVE.
After 20 years, the value is $121,791.14.

9. N = 12(12) = 144, I% = 5.5, PV = 0, PMT = 450, temporarily skip FV, P/Y and C/Y = 12, BEGIN.
Move the cursor to FV and press ALPHA SOLVE.
After 12 years, the value is $91,912.73.

11. Amount financed 170,000 – 50,000 = 120,000
N = 10(12) = 120, I% = 7.25, PV = 120,000, temporarily skip PMT, FV = 0, P/Y and C/Y = 12, END. Move the cursor to PMT and press ALPHA SOLVE.
The monthly payment is $1408.81.

12. N = 5(12) = 60, I% = 6.875, PV = 8500, temporarily skip PMT, FV = 0, P/Y and C/Y = 12, END. Move the cursor to PMT and press ALPHA SOLVE.
The monthly payment is $167.81.

13. N = 5(12) = 60, I% = 4.75, temporarily skip PV, PMT = 5000, FV = 0, P/Y and C/Y = 12, BEGIN.
Move the cursor to PV and press ALPHA SOLVE.
The present value is $267,623.57.

15. N = 15(2) = 30, I% = 6, PV = 0, temporarily skip PMT, FV = 10,000,000, P/Y and C/Y = 2, END. Move the cursor to PMT and press ALPHA SOLVE.
The semi-annual payment is $210,192.59.

17. N = 3(12) = 36, I% = 5, PV = 0, temporarily skip PMT, FV = 15,000, P/Y and C/Y = 12, BEGIN.
Move the cursor to PMT and press ALPHA SOLVE.
The monthly payment should be $385.46.

THINK ABOUT IT, SECTION 4.4

1. $\dfrac{18\%}{12 \text{ months}} = 1.5\%$ is the monthly interest rate.

2. $200 + $10 = $210

EXERCISE SET 4.4

1. $0.045 \cdot 18{,}456.32 = 830.5344 \approx 830.53$
The greater of $830.53 and $500 is $830.53. The minimum payment is $830.53.

3. Recording the balances:

Date	Payment or Purchases ($)	Balance Each Day ($)	Days	Unpaid Balance Times Number of Days ($)
Apr. 14		653		
Apr. 19	155.56	653+155.56=808.56	5	$5 \cdot 653 + 155.56 = 3420.56$
Apr. 21	22.45	808.56+22.45=831.01	2	$2 \cdot 808.56 + 22.45 = 1639.57$
Apr. 25	–500	831.01–500=331.01	4	$4 \cdot 831.01 - 500 = 2824.04$
Apr 28	201.36	331.01+201.36=532.37	3	$3 \cdot 331.01 + 201.36 = 1194.39$
			16	$16 \cdot 532.37 = 8517.92$
Total				17,596.48

Average daily balance $= \dfrac{17{,}596.48}{30} \approx 586.5493$ The average daily balance is $586.55.

5. Recording the balances:

Date	Payment or Purchases ($)	Balance Each Day ($)	Days	Unpaid Balance Times Number of Days ($)
Sept. 14		562		
Sept. 19	286	562+286=848	5	$5 \cdot 562 + 286 = 3096$
Sept. 25	–350	848–350=498	6	$6 \cdot 848 - 350 = 4738$
Sept. 28	15	498+15=513	3	$3 \cdot 498 + 15 = 1509$
			16	$16 \cdot 513 = 8208$
Total				17,551

Average daily balance $= \dfrac{17{,}551}{30} \approx 585.03$ The average daily balance is $585.03.

Finance charge $= 585.03 \cdot \dfrac{0.205}{365} \cdot 30 \approx 9.86$ The finance charge is $9.86.

7. Using a graphing calculator,
temporarily skip N, I% = 19.6, PV = 7841.36, PMT = –500, FV = 0, P/Y and C/Y = 12, END. Move the cursor to N and press ALPHA SOLVE. N ≈ 18.2649
It will take 19 months to repay the credit card debt.

9. Using the spreadsheet Average Daily Balance

Date payment is due	7/15/18
Date of next payment	8/15/18
Days in billing cycle	31
Annual Interest Rate	20.50%
Beginning Balance	856.25

- Enter the payment due date
- Enter the next payment due date
- Enter the annual interest rate
- Enter the beginning balance
- Enter the date of a transaction and its amount
- Do not enter values in any shaded cell.

Date of Next Transaction	Purchase or Payment ($)	Balance ($)	Number of Days until Balance Changes	Unpaid Balance times Number of Days ($)
7/15/18		856.25		
7/19/18	115.59	971.84	4	$3,540.59
7/24/18	29.19	1,001.03	5	$4,888.39
7/25/18	110.56	1,111.59	1	$1,111.59
7/26/18	36.07	1,147.66	1	$1,147.66
7/26/18	53.59	1,201.25	1	$53.59
7/27/18	86.46	1,287.71	1	$1,287.71
7/29/18	-450.00	837.71	2	$2,125.42
7/30/18	29.12	866.83	1	$866.83
8/4/18	73.01	939.84	5	$4,407.16
8/5/18	21.62	961.46	1	$961.46
8/6/18	126.99	1,088.45	1	$1,088.45
8/7/18	59.95	1,148.40	1	$1,148.40
8/7/18	107.55	1,255.95	1	$107.55
8/8/18	141.30	1,397.25	1	$1,397.25
8/8/18	51.91	1,449.16	1	$51.91
8/9/18	142.09	1,591.25	1	$1,591.25
8/10/18	48.14	1,639.39	1	$1,639.39
8/11/18	85.65	1,725.04	1	$1,725.04
8/13/18	144.49	1,869.53	2	$3,594.57
8/14/18	105.58	1,975.11	1	$1,975.11
			1	$1,975.11
			Total	$36,684.43
			Average daily balance	$1,183.37
			Finance charge	$20.60

The finance charge is $20.60.

11. Using a graphing calculator,
N = 2(12) = 24, I% = 19.6, PV = 3540.85, temporarily skip PMT, FV = 0, P/Y and C/Y = 12, END. Move the cursor to PMT and press ALPHA SOLVE.
The monthly payment is $179.52.

13. a. Using a graphing calculator,
N = 15, I% = 21.3, PV = –8000, PMT = 400, temporarily skip FV = 0, P/Y and C/Y = 12, END. Move the cursor to FV and press ALPHA SOLVE.
$FV \approx 3610.1186$
After 15 months, Jon will owe $3610.12.

b. Make 15 payments of $400 each.
$8000 - 400(15) = 2000$
After 15 months, the balance is $2000.

c. $3610.12 - 2000 = 1610.12$
The balance is $1610.12 lower on the School Credit Union Card.

15. Using a graphing calculator,
N = 12, I% = 21.5, PV = 1576.57, PMT = 0, temporarily skip FV = 0, P/Y and C/Y = 12, END. Move the cursor to FV and press ALPHA SOLVE. $FV \approx 1951.0121$
After 1 year, Richard will owe $1951.01.

17. a. $0.02 \cdot 2500 = 50$
The greater of $50 and $25 is $50.
The minimum payment is $50.

Using a graphing calculator,
N = 1, I% = 25, PV = –2500 + 50 = –2450, PMT = 0, temporarily skip FV = 0, P/Y and C/Y = 12, END.
Move the cursor to FV and press ALPHA SOLVE. $FV \approx 2501.0417$

The credit card balance will be $2501.04.

b. By making the minimum payment, your debt is increasing even though you are not using the card.

THINK ABOUT IT, SECTION 4.5

1. For a subsidized student loan, the government pays the interest on the loan while the student is in school. For a nonsubsidized loan, the student must pay the accrued interest on the loan.

2. With the nonsubsidized loan, the interest begins accruing while the student is in school. With the subsidized loan, the interest begins accruing after graduation. The student will pay more interest on the nonsubsidized loan.

3. The discount fee is an amount subtracted from the loan amount. It is usually a percent of the loan amount.

EXERCISE SET 4.5

1. Loan fee $= 0.0101 \cdot 7250 = 73.225$
Amount received $= 7250 - 73.225 = 7176.775$
The student receives $7176.78.

3. Since this is a nonsubsidized loan, interest accrues during the period of time when payments on the loan are not being made.
The interest is
$$I = Prt$$
$$= 18{,}000(0.053)4 = 3816$$
Amount owed $= 18{,}000 + 3816 = 21{,}816$

N = 12(12) = 144, I% = 5.3, PV = 21,816, temporarily skip PMT, FV = 0, P/Y and C/Y = 12, END.
Move the cursor to PMT and press ALPHA SOLVE.
The monthly payment is $205.07.

5. Since this is a subsidized loan, interest does not accrue during the period of time when payments on the loan are not being made.

N = 10(12) = 120, I% = 4.125, PV = 21,000, temporarily skip PMT, FV = 0, P/Y and C/Y = 12, END.
Move the cursor to PMT and press ALPHA SOLVE.
The monthly payment is $213.86.

7. Temporarily skip N, I% = 4, PV = 42,000, PMT = 425, FV = 0, P/Y and C/Y = 12, END.
Move the cursor to N and press ALPHA SOLVE.
It will take approximately 120 months to repay the loan.

9. Total loan amount = 4(5000) = $20,000

Since this is a subsidized loan, interest does not accrue during the period of time when payments on the loan are not being made.

N = 10(12) = 120, I% = 3.75, PV = 20,000, temporarily skip PMT, FV = 0, P/Y and C/Y = 12, END.
Move the cursor to PMT and press ALPHA SOLVE.
The monthly payment is $200.12.

11. For the nonsubsidized loan
$$I = Prt$$
$$= 28,000(0.035)2 = 1960$$
Amount owed = 28,000 + 1960 = 29,960

N = 8(12) = 96, I% = 3.5, PV = 29,960, temporarily skip PMT, FV = 0, P/Y and C/Y = 12, END.
Move the cursor to PMT and press ALPHA SOLVE.
The monthly payment is $358.26.
Total amount paid: 96(358.26) = 34,392.96.
Total interest paid: 34,392.96 – 28,000
= 6392.96

For the subsidized loan
N = 8(12) = 96, I% = 4.5, PV = 28,000, temporarily skip PMT, FV = 0, P/Y and C/Y = 12, END.
Move the cursor to PMT and press ALPHA SOLVE.
The monthly payment is $347.85.
Total amount paid: 96(347.85) = 33,393.60.
Total interest paid: 33,393.60 – 28,000
= 5393.60
Katrina pays less interest for the subsidized loan. The interest paid on the nonsubsidized loan is $6392.96. The interest paid on the subsidized loan is $5393.60.

13. Deferred interest $= Prt$
$$= 50,000 \cdot 0.0578 \cdot 4 = 11,560$$
Amount owed = 50,000 + 11,560 = 61,560

N = 10(12) = 120, I% = 5.78, PV = 61,560, temporarily skip PMT, FV = 0, P/Y and C/Y = 12, END.
Move the cursor to PMT and press ALPHA SOLVE.
The monthly payment is $676.66.

15. The monthly payment in Exercise 13 is $676.66.
Student receives $= 50,000 - 50,000 \cdot 0.0476$
$$= 47,620$$
N = 10(12) = 120, temporarily skip I%, PV = 47,620, PMT = 676.66, FV = 0, P/Y and C/Y = 12, END.
Move the cursor to I% and press ALPHA SOLVE.
The APR is 11.76%.

17. For the $25,000 loan for 12 years,
N = 12(12) = 144, I% = 6, PV = 25,000, temporarily skip PMT, FV = 0, P/Y and C/Y = 12, END.
Move the cursor to PMT and press ALPHA SOLVE.
The monthly payment is $243.96.
Total amount paid: 144(243.96) = 35,130.24
Total interest paid: 35,130.24 – 25,000
= 10,130.24

For the $15,000 loan for 8 years,
N = 8(12) = 96, I% = 7, PV = 15,000, temporarily skip PMT, FV = 0, P/Y and C/Y = 12, END.
Move the cursor to PMT and press ALPHA SOLVE.
The monthly payment is $204.51.
Total amount paid: 96(204.51) = 19,632.96
Total interest paid: 19,632.96 – 15,000
= 4632.96

For the $40,000 loan for 10 years,
N = 10(12) = 120, I% = 6.5, PV = 40,000, temporarily skip PMT, FV = 0, P/Y and C/Y = 12, END.
Move the cursor to PMT and press ALPHA SOLVE.
The monthly payment is $454.19.
Total amount paid: 120(454.19) = 54,502.80.
Total interest paid: 54,502.80 – 40,000
= 14,502.80

Yes, he should take the consolidation offer. The total interest on the two loans is $10,130.24 + $4632.96 = $14,763.20. The total interest on the 10-year loan is $14,502.80.

THINK ABOUT IT, SECTION 4.6

1. No. The car depreciates by 20% of its current value, not 20% of its original value.

2. 100/4 > 100/5
4 gallons per 100 miles has the best miles per gallon rating.

3. 45% of residual value < 50% of residual value
It is better for the residual value to be 45% of the cost.

EXERCISE SET 4.6

1. For the $53,976.47 loan for 5 years,
N = 5(12) = 60, I% = 5.1, PV = 53,976.47, temporarily skip PMT, FV = 0, P/Y and C/Y = 12, END.
Move the cursor to PMT and press ALPHA SOLVE.
The monthly payment is $1021.08.

3. $1000 \text{ miles} \cdot \dfrac{1 \text{ gallon}}{23 \text{ miles}} \approx 43.478 \text{ gallons}$

$(43.478 \text{ gallons})\left(\dfrac{\$3.06}{1 \text{ gallon}}\right) = \$133.04$

The fuel cost for 1 month is $133.04.

5. Sales tax = 22,350(0.0625) = 1396.88
Total cost = 22,350 + 1396.88 + 452
= 24,198.88
Down payment = 24,198.88(0.10) = 2419.89
Amount financed
= Cost – down payment
= 24,198.88 – 2419.89 = 21,778.99
For the $21,778.99 loan for 4 years,
N = 4(12) = 48, I% = 4.9, PV = 21,778.99, temporarily skip PMT, FV = 0, P/Y and C/Y = 12, END.
Move the cursor to PMT and press ALPHA SOLVE.
The monthly payment is $500.57.

7. There are 4(12) – 30 = 18 payments remaining.
N = 18, I% = 4.95, temporarily skip PV, PMT = 1472.43, FV = 0, P/Y and C/Y = 12, END.
Move the cursor to PV and press ALPHA SOLVE.
The orthodontist still owes $25,493.08 on the car.

9. N = 5(12) = 60, I% = 2.75, temporarily skip PV, PMT = –475, FV = 0, P/Y and C/Y = 12, END.
Move the cursor to PV and press ALPHA SOLVE.
The maximum she can finance is $26,599.

11. $\dfrac{35,600}{1000} \cdot 18.416522 \approx 655.63$

The monthly payment is $655.63.

13. a. $28(0.30) = 8.4$

$28 - 8.4 = 19.6 \approx 20$

The gas mileage is approximately 20 mpg.

b. Cost for driving 50 miles with 28 mpg:

$\dfrac{50 \text{ miles}}{28 \text{ miles/gallon}} \cdot \dfrac{\$2.89}{1 \text{ gallon}} \approx 5.16$

Cost for driving 50 miles with 20 mpg:

$\dfrac{50 \text{ miles}}{20 \text{ miles/gallon}} \cdot \dfrac{\$2.89}{1 \text{ gallon}} \approx 7.23$

Cost difference: 7.23 – 5.16 = 2.07.
The increased cost is about $2.

15. $48,357(0.22) = 10,638.54$

The depreciation is $10,638.54.

$\dfrac{\$10,638.54}{12,470 \text{ miles}} = \dfrac{x \text{ dollars}}{100 \text{ miles}}$

$x \approx 85.31$

The depreciation is $85.31 per 100 miles driven.

17. Net cost = 41,250 – 41,250(0.20)
= 33,000
Residual value = 45,900(0.52) = 23,868

$F = (33,000 + 23,868)0.0026$
$= 147.8568$

Average monthly finance charge is $147.86.

$D = \dfrac{33,000 - 23,868}{24}$
$= 380.50$

Average monthly depreciation is $380.50.

$P = D + F = 380.50 + 147.86$
$= 528.36$

The monthly lease payment is $528.36.

19. There are 4(12) – 28 = 20 payments remaining.
N = 20, I% = 5.5, temporarily skip PV, PMT = 976.77, FV = 0, P/Y and C/Y = 12, END.
Move the cursor to PV and press ALPHA SOLVE.
The payoff is $18,626.04.

21. First find the first year interest paid:
Find the payoff amount after 1 year.
N = 48 – 12 = 36, I% = 4.3, temporarily skip PV, PMT = 567.84
Solving for PV = 19,146.51
Amount of principal paid:
25,000 – 19,146.51 = 5853.49
Total payments: 12(567.84) = 6814.08
Total interest paid in year 1:
6814.08 – 5853.49 = 960.59

Find the total cost:
License and registration:
0.75% of purchase price
= 29,900(0.0075) = 224.25
Sales tax:
6.25% of purchase price
= 29,900(0.0625) = 1868.75
Depreciation:
21% of purchase price
= 29,900(0.21) = 6279.00
Fuel cost driving 14,000 miles:
Average cost of gasoline $2.74 per gallon

$\dfrac{14,000 \text{ miles}}{28 \text{ miles/gallon}} \cdot \left(\dfrac{\$2.74}{\text{gallon}}\right) = 1370.00$

Year 1 total interest paid: = 960.59
Insurance: = 712.00
Maintenance: = 356.00
Total = $11,770.59

23. For 20 mpg to 30 mpg, a 10 mpg increase

$$\frac{10,000 \text{ miles}}{20 \text{ miles/gallon}} \cdot \left(\frac{\$3.12}{\text{gallon}}\right) = 1560$$

$$\frac{10,000 \text{ miles}}{30 \text{ miles/gallon}} \cdot \left(\frac{\$3.12}{\text{gallon}}\right) = 1040$$

A savings of 1560 – 1040, or $520.

For 35 mpg to 50 mpg, a 15 mpg increase

$$\frac{10,000 \text{ miles}}{35 \text{ miles/gallon}} \cdot \left(\frac{\$3.12}{\text{gallon}}\right) \approx 891.43$$

$$\frac{10,000 \text{ miles}}{50 \text{ miles/gallon}} \cdot \left(\frac{\$3.12}{\text{gallon}}\right) = 624$$

A savings of 891.43 – 624, or $267.43.
Thus, the person going from 20 mpg to 30 mpg has the greater savings.

25. Amount financed = 5606.77 + 28,905.87
= 34,512.64
N = 5(12) = 60, I% = 4.25, PV = 34,512.64, temporarily skip PMT, FV = 0, P/Y and C/Y = 12, END.
Move the cursor to PMT and press ALPHA SOLVE.
The monthly payment is $639.50.

27. Net cost
= 62,500 – 25,000 – 5000
= 32,500
Residual value = 65,950(0.45) = 29,677.50

$$F = (32,500 + 29,677.50)\frac{4.5}{2400} \approx 116.58$$

Average monthly finance charge is $116.58.

$$D = \frac{32,500 - 29,677.50}{36} \approx 78.40$$

Average monthly depreciation is $78.40.

$$P = D + F = 78.40 + 116.58 = 194.98$$

The monthly lease payment without sales tax is $194.98.
Adding sales tax
= 194.98 + 194.98(0.0725) ≈ 209.17
The monthly lease payment with sales tax is $209.17.

29. a. For 32 mpg to 36 mpg, a 4 mpg increase

$$\frac{10,000 \text{ miles}}{32 \text{ miles/gallon}} \cdot \left(\frac{\$3.41}{\text{gallon}}\right) = 1065.63$$

$$\frac{10,000 \text{ miles}}{36 \text{ miles/gallon}} \cdot \left(\frac{\$3.41}{\text{gallon}}\right) \approx 947.22$$

A savings of 1065.63 – 947.22, or $118.41.

b. N = 5, I% = 4.2, PV = 0, PMT = 118.41,
Solving for FV = 643.92
A savings of $643.92.

31. In Exercise 11, the monthly payment is $655.63 on $35,600 for a 5-year loan.
N = 5(12) = 60, temporarily skip I%,
PV = 35,600, PMT = –655.63, FV = 0, P/Y and C/Y = 12, END.
Move the cursor to I% and press ALPHA SOLVE.
The interest rate is 4%.

33. Cost for battery alone:
Miles: 0.40(12,000) = 4800

$$4800 \text{ miles} \cdot \frac{39 \text{ kWh}}{100 \text{ miles}} \cdot \left(\frac{\$0.18}{\text{kWh}}\right) = 336.96$$

Cost for gasoline:
Miles: 0.60(12,000) = 7200

$$\frac{7200 \text{ miles}}{33 \text{ miles/gallon}} \cdot \left(\frac{\$3.10}{\text{gallon}}\right) \approx 676.36$$

Thus, the total cost is 336.96 + 676.36, or $1013.32.

THINK ABOUT IT, SECTION 4.7

1. No. The DTI ratio must be less than 35.

2. The interest rate for a fixed-rate mortgage remains the same throughout the term of the loan. The interest rate for an adjustable-rate mortgage changes on a periodic basis.

3. In March, $500 goes to paying principal. In April, since the principal is $500 less, then the payment will be less than $700 for interest and MORE than $500 for principal.

4. 1.5%

EXERCISE SET 4.7

1. $\frac{250 + 364.97 + 187 + 2784.64}{5630} \approx 0.6371$
The debt-to-income ratio is approximately 64%.

3. a. Commission: 490,000(0.06) = 29,400
Title insurance: 490,000(0.012) = 5880
Total fees: 29,400 + 5880 + 875 = 36,155
Amount: 490,000 – 36,155 = 453,845
After fees, they receive $453,845.00.

b. $\frac{36,155}{490,000} \approx 0.07379$, or about 7.4%

5. 322,000(0.0125) = 4025
The discount points are $4025.

7. 410,000(0.008) = 3280
The PMI adds $3280 to the cost.

9. Using a graphing calculator,
N = 12(30) = 360, I% = 4.25, PV = 295,000, temporarily skip PMT, FV = 0, P/Y and C/Y = 12. Solving for PMT = 1451.22
The monthly mortgage payment is $1451.22.

11. Using a graphing calculator,
N = 360 – 60 = 300, I% = 6.5, temporarily skip PV, PMT = –1673.05, FV = 0, P/Y and C/Y = 12. Solving for PV = 247,783.21

N = 12(30) = 360, I% = 4.5, PV = 247,783.21, temporarily skip PMT, FV = 0, P/Y and C/Y = 12. Solving for PMT = 1255.48
The new monthly payment will be $1255.48.

13. N = 360 – 84 = 276, I% = 3, temporarily skip PV, PMT = –2635.03, FV = 0, P/Y and C/Y = 12. Solving for PV = 524,889.39

N = 276, I% = 3.5, PV = 524,889.39, temporarily skip PMT, FV = 0, P/Y and C/Y = 12. Solving for PMT = 2771.47
The new monthly payment will be $2771.47.

15. Difference in cost of points:
350,000(0.0185) – 350,000(0.015) = 1225

Monthly mortgage payment for 4.1%
N = 12(30) = 360, I% = 4.1, PV = 350,000, temporarily skip PMT, FV = 0, P/Y and C/Y = 12. Solving for PMT = 1691.19
The monthly mortgage payment is $1691.19.

Monthly mortgage payment for 3.8%
N = 12(30) = 360, I% = 3.8, PV = 350,000, temporarily skip PMT, FV = 0, P/Y and C/Y = 12. Solving for PMT = 1630.85
The monthly mortgage payment is $1630.85.

The difference in payments is
1691.19 – 1630.85 = 60.34

Months to recoup costs $\frac{1225}{60.34} \approx 20.302$
It will take about 21 months to recoup the cost.

17. Fee for points $= 675,000 \cdot 0.0125 = 8437.50$
Loan origination fee $= 675,000 \cdot 0.005 = 3375$
Adjusted loan balance
= 675,000 + 8437.50 + 3375 = $686,812.50

Find the monthly payment.
N = 12(25) = 300, I% = 3.625, PV = 686,812.50, temporarily skip PMT, FV = 0, P/Y and C/Y = 12.
Solving for PMT = 3484.56
The monthly payment is $3484.56.

Change PV from 686,812.50 to 675,000.
N = 300, temporarily skip I%, PV = 675,000, PMT = 3484.56, FV = 0, P/Y and C/Y = 12. Solving for I% = 3.7885
The APR for the loan is 3.789%.

19. a. For the 30-year loan
N = 360, I% = 4.25, PV = 375,000, temporarily skip PMT, FV = 0, P/Y and C/Y = 12. Solving for PMT = 1844.77
The monthly payment is $1844.77.

For the 25-year loan
N = 12(25) = 300, I% = 4.25, PV = 375,000, temporarily skip PMT, FV = 0, P/Y and C/Y = 12.
Solving for PMT = 2031.52
The monthly payment is $2031.52.

b. For 30-year loan, total interest is
360(1844.77) – 375,000 = 289,117.20
For the 25-year loan, total interest is
300(2031.52) – 375,000 = 234,456
The savings in interest is
289,117.20 – 234,456, or $54,661.20.

21. a. For the 30-year loan
N = 12(30) = 360, I% = 3.95, PV = 415,000, temporarily skip PMT, FV = 0, P/Y and C/Y = 12.
Solving for PMT = 1969.33
The monthly payment is $1969.33.

For the 15-year loan
N = 12(15) = 180, I% = 3.95, PV = 415,000, temporarily skip PMT, FV = 0, P/Y and C/Y = 12.
Solving for PMT = 3059.32
The monthly payment is $3059.32.

b. For 30-year loan, total interest is
360(1969.33) – 415,000 = 293,958.80
For the 15-year loan, total interest is
180(3059.32) – 415,000 = 135,677.60
The savings in interest is
293,958.80 – 135,677.60, or $158,281.20.

c. $\frac{1969.33}{3059.32} \approx 0.6437$
No, it is about 64% of the 15-year loan payment.

d. $\frac{158,281.20}{293,958.80} \approx 0.5384$
The interest savings on the 15-year loan is more than one-half of the interest paid on the 30-year loan.

23. Use the spreadsheet Amortization Schedule with Extra Payments
    a. To find when the amount of principal in a monthly payment exceeds the amount of interest, compare columns E and F. In month 176, amount to principal is \$1265.82 and amount to interest is \$1267.61.
    b. From the spreadsheet: Total interest paid is \$412,028.22.
    c. No. It is the same month. Change loan amount to \$750,000. In month 176, principal is \$1898.74 and interest is \$1901.40.

THINK ABOUT IT, SECTION 4.8

1. Use the formula $I = Prt$, where $I = 2.50$ and $t = 1$ and solve for r.
   $$I = Prt$$
   $$2.50 = Pr(1)$$
   $$\frac{2.50}{P} = r$$
   By increasing the stock's value, P, the dividend yield will go down.

2. $$\text{P/E} = \frac{\text{Stock price}}{\text{EPS}}$$
   By increasing EPS while the stock price remains the same, the P/E will decrease.

3. Greater than

4. Discount

5. Real rate of return considers the effects of inflation on the value of an asset.

EXERCISE SET 4.8

1. $$\text{Real rate of return} = \frac{1+\text{nominal rate}}{1+\text{inflation rate}} - 1 = \frac{1+0.034}{1+0.012} - 1 \approx 0.021739$$
   The real rate of return is 2.17%.

3. Use the formula $I = Prt$, where $I = 1.27$, $P = 41.05$, and $t = 1$.
   $$I = Prt$$
   $$1.27 = 41.05r(1)$$
   $$r \approx 0.03094$$
   The dividend yield is 3.1%.

5. Net asset value of one share
   $$= \frac{\$10,250,000,000}{900,000,000} \approx 11.39$$
   The net asset value of one share is \$11.39.

7. Use the formula $I = Prt$, where $P = 5000$, $r = 0.055$, and $t = 7$.
   $$I = Prt$$
   $$I = 5000(0.055)(7)$$
   $$I = 1925$$
   The total interest payments are \$1925.

9. Earnings = \$2.2 billion, number of shares = 1.02 billion
   $$\text{EPS} = \frac{2,200,000,000}{1,020,000,000} \approx 2.16$$
   The EPS for the company is \$2.16.

11. EPS = \$13.89, stock price = \$167.13
    $$\text{P/E} = \frac{\text{Stock price}}{\text{EPS}} = \frac{167.13}{13.89} \approx 12.0$$
    The P/E is 12.0.

13. Interest on 4% bond = Interest on 3.5% bond
    $$1000(0.04) = x(0.035)$$
    $$40 = 0.035x$$
    $$x \approx 1143$$
    The nominal value of the bond is \$1143.

15. Dividend per share = \$1.966, Number of shares = 500
    $$\text{Dividends} = 1.966(500) = 983$$
    The shareholder earned \$983 in dividends.

17. $\text{Fee} = 0.0625 \cdot \$10,000 = \$625$
    Funds available = \$10,000 – \$625 = \$9375
    Number of shares purchased
    $$= \frac{\$9375}{\$42.08} \approx 222.7899$$
    Seth can purchase 222.790 shares.

19. $$\text{Fee} = 0.0024 \cdot \$11,300,000,000 = \$27,120,000$$
    Funds available
    $$= \$11,300,000,000 - \$27,120,000 = \$11,272,880,000$$
    Net asset value of one share
    $$= \frac{\$11,272,880,000}{750,000,000} \approx 15.03$$
    The net asset value of one share is \$15.03.

21. EPS = \$6.04, stock price = \$146.92
    $$\text{P/E} = \frac{\text{Stock price}}{\text{EPS}} = \frac{146.92}{6.04} \approx 24.3$$
    The P/E is 24.3.

23. Dividend per share = \$0.66,
Number of shares = 450
$\text{Dividends} = 0.66(450) = 297$
The shareholder earned \$297 in dividends.

25. Earnings = \$7.1 billion,
number of shares = 1.3 billion
$\text{EPS} = \dfrac{7,100,000,000}{1,300,000,000} \approx 5.46$
The EPS for the company is \$5.46.

27. Use the formula $I = Prt$, where $I = 3.08$, $P = 119.30$, and $t = 1$.
$I = Prt$
$3.08 = 119.30r(1)$
$r \approx 0.02582$
The dividend yield is 2.58%.

CHAPTER 4 REVIEW EXERCISES

1. a. The interest rate is per year and the time period is in years.
$I = Prt = 2750(0.0675)(3) \approx 556.88$
The interest due is \$556.88.

   b. The interest rate is per month and the time period is in months.
$I = Prt = 2750(0.009)(3) = 74.25$
The interest due is \$74.25.

2. Because the interest rate is an annual rate, the time must be measured in years:
$t = \dfrac{8 \text{ months}}{1 \text{ year}} = \dfrac{8 \text{ months}}{12 \text{ months}} = \dfrac{8}{12}$
$I = Prt = 8500(0.09)\left(\dfrac{8}{12}\right) = 510$
The interest earned is \$510.

3. The value for the time is
$I = Prt$
$25 = 500r\left(\dfrac{14}{365}\right)$
$r \approx 1.30357$
The interest rate is approximately 130.4%.

4. Since the interest rate is per year, the time period of the loan is $t = \dfrac{108}{365}$ year.
$A = P(1+rt) = 7000\left[1+0.104\left(\dfrac{108}{365}\right)\right]$
≈ 7215.41
The maturity value of the loan is \$7215.41.

5. Solve the simple interest formula for r.
$I = Prt$
$127.50 = 6800r\left(\dfrac{3}{12}\right)$
$127.50 = 1700r$
$\dfrac{127.50}{1700} = r$
$r = 7.5\%$
The simple interest rate is 7.5%.

6. Solve the maturity value formula for r.
$A = P + Prt$
$5000 = 4950 + 4950r\left(\dfrac{3}{12}\right)$
$5000 = 4950 + 1237.5r$
$50 = 1237.5r$
$\dfrac{50}{1237.5} = r$
$r \approx 0.0404$
The simple interest rate is 4.0%.

7. $P = 3000, r = 6.6\% = 0.066, n = 12, t = 3$
$A = P\left(1+\dfrac{r}{n}\right)^{nt}$
$= 3000\left(1+\dfrac{0.066}{12}\right)^{(12\cdot 3)}$
$= 3000(1.0055)^{36} \approx 3654.90$
The compound amount after 3 years is approximately \$3654.90.

8. $P = 6400, r = 6\% = 0.06, n = 4, t = 10$
$A = P\left(1+\dfrac{r}{n}\right)^{nt}$
$= 6400\left(1+\dfrac{0.06}{4}\right)^{(4\cdot 10)}$
$= 6400(1.015)^{40} \approx 11,609.72$
The compound amount after 10 years is approximately \$11,609.72.

9. $P = 6000, r = 9\% = 0.09, n = 365, t = 3$
$A = P\left(1+\dfrac{r}{n}\right)^{nt}$
$= 6000\left(1+\dfrac{0.09}{365}\right)^{(365\cdot 3)}$
≈ 7859.53
The future value is approximately \$7859.53.

10. $P = 600, r = 7.2\% = 0.072, n = 365, t = 4$
$A = P\left(1+\dfrac{r}{n}\right)^{nt}$
$= 600\left(1+\dfrac{0.072}{365}\right)^{(365\cdot 4)}$
≈ 800.23
Calculate the interest earned.
$I = A - P = 800.23 - 600 = 200.23$
The amount of interest earned is \$200.23.

11. $A = 18{,}500$, $r = 8\% = 0.08$, $n = 2$, $t = 7$

$$P = \frac{A}{\left(1+\frac{r}{n}\right)^{nt}} = \frac{18{,}500}{\left(1+\frac{0.08}{2}\right)^{(2\cdot 7)}} = \frac{18{,}500}{(1.04)^{14}} \approx 10{,}683.29$$

$10,683.29 should be invested in the account in order to have $18,500 in 7 years.

12. $A = 80{,}000$, $r = 8\% = 0.08$, $n = 4$, $t = 18$

$$P = \frac{A}{\left(1+\frac{r}{n}\right)^{nt}} = \frac{80{,}000}{\left(1+\frac{0.08}{4}\right)^{(4\cdot 18)}} = \frac{80{,}000}{(1.02)^{72}} \approx 19{,}225.50$$

$19,225.50 should be invested in the account in order to have $80,000 in 18 years.

13. $A = 250{,}000$, $r = 3.1\% = 0.031$ and $t = 10$. The inflation rate is an annual rate, so $n = 1$.

$$P = \frac{A}{\left(1+\frac{r}{n}\right)^{nt}} = \frac{250{,}000}{\left(1+\frac{0.031}{1}\right)^{(1\cdot 10)}} = \frac{250{,}000}{(1.031)^{10}} \approx 184{,}227.03$$

The purchasing power of the $250,000 will be approximately $184,227.03 in 10 years.

14. $$\text{Inflation rate} = 100\left(\frac{\text{new CPI} - \text{old CPI}}{\text{old CPI}}\right) = 100\left(\frac{231.258 - 228.936}{228.936}\right) \approx 1.0143$$

The inflation rate is approximately 1.0%.

15. $$275{,}000 \cdot \frac{274.561}{264.257} \approx 285{,}722.89$$

An equivalent house would cost $285,722.89 in 2018.

16. $$59{,}000 \cdot \frac{245.168}{229.594} \approx 63{,}002.13$$

No, but it is only shy by about $2.

17. $n = 12$, APR = 8% = 0.08

$$\text{APY} = \left(1+\frac{\text{APR}}{n}\right)^{n} - 1 = \left(1+\frac{0.08}{12}\right)^{12} - 1 \approx 0.0829995$$

The APY is approximately 8.3%.

18. Calculate $\left(1+\frac{r}{n}\right)^{nt}$ for each investment. 5.2% compounded quarterly: $n = 4$, $r = 5.2\% = 0.052$, $t = 1$

$$\left(1+\frac{r}{n}\right)^{nt} = \left(1+\frac{0.052}{4}\right)^{(4\cdot 1)} \approx 1.053022817$$

5.4% compounded semiannually: $n = 2$, $r = 5.4\% = 0.054$, $t = 1$

$$\left(1+\frac{r}{n}\right)^{nt} = \left(1+\frac{0.054}{2}\right)^{(2\cdot 1)} = 1.054729$$

Compare the two compound amounts.
$1.054729 > 1.053022817$
So 5.4% compounded semiannually has a higher yield than 5.2% compounded quarterly.

19. Calculate the future value based on monthly payments of $250.
N = 20(12) = 240, I% = 4.5, PMT = 250, PV = 0, temporarily skip FV, P/Y and C/Y = 12, BEGIN. Move the cursor to FV and press ALPHA SOLVE.
The future value is $97,394.96

20. N = 15(52) = 780, I% = 5.25, PV = 0, PMT = 50, temporarily skip FV, P/Y and C/Y = 52, END.
Move the cursor to FV and press ALPHA SOLVE.
After 15 years, the value is $59,281.08.

21. N = 3(12) = 36, I% = 3.125, PV = 0, temporarily skip PMT, FV = 8000, P/Y and C/Y = 12, BEGIN. Move the cursor to PMT and press ALPHA SOLVE.
The monthly payment is $211.71.

22. N = 25(2) = 50, I% = 6.5, PV = 0, temporarily skip PMT, FV = 50,000,000, P/Y and C/Y = 2, END. Move the cursor to PMT and press ALPHA SOLVE.
The semi-annual payment is $411,513.73.

23. N = 10(12) = 120, I% = 4.875, temporarily skip PV, PMT = 2500, FV = 0, P/Y and C/Y = 12, END.
Move the cursor to PV and press ALPHA SOLVE.
The present value is $237,066.65.

24. N = 8(12) = 96, I% = 4.75, temporarily skip PV, PMT = 1250, FV = 0, P/Y and C/Y = 12, BEGIN.
Move the cursor to PV and press ALPHA SOLVE.
The present value is $100,065.72.

25. Find the present value of the monthly payment.
N = 20(12) = 240, I% = 4.25, temporarily skip PV, PMT = 25,000, FV = 0, P/Y and C/Y = 12, END.
Move the cursor to PV and press ALPHA SOLVE.
The present value is $4,037,242.96, which is much greater than $1,000,000.

Therefore, $25,000 per month for 20 years has the greater present value.

26. Down payment = 8500(0.10) = 850
Amount financed = 8500 – 850 = 7650
N = 3(12) = 36, I% = 5.5, PV = 7650, temporarily skip PMT, FV = 0, P/Y and C/Y = 12, END.
Move the cursor to PMT and press ALPHA SOLVE.
The monthly payment is $231.00.

27. Down payment = 45,000(0.25) = 11,250
Amount financed = 45,000 – 11,250 = 33,750
N = 8(12) = 96, I% = 3.8, PV = 33,750, temporarily skip PMT, FV = 0, P/Y and C/Y = 12, END.
Move the cursor to PMT and press ALPHA SOLVE.
The monthly payment is $408.26.

Total paid = 408.26(96) = 39,192.96
Amount of interest
= 39,192.96 – 33,750 = 5442.96
The total amount of interest paid is $5442.96.

28. Since Dasan has made 3(12) = 36 payments, N = 60 – 36 = 24. Then I% = 4.8, temporarily skip PV, PMT = –650.80, FV = 0, P/Y and C/Y = 12, END.
Move the cursor to PV and press ALPHA SOLVE.
Dasan's payoff is $14,864.60.

29. N = 2(12) = 24, I% = 8.5, PV = 999, temporarily skip PMT, FV = 0, P/Y and C/Y = 12, END.
Move the cursor to PMT and press ALPHA SOLVE.
The monthly payment is $45.41.

30. Recording the balances:

Date	Payment or Purchases ($)	Balance Each Day ($)	Days	Unpaid Balance Times Number of Days ($)
Mar. 11		423.35		
Mar. 18	145.50	423.35+145.50=568.85	7	$7 \cdot 423.35 + 145.50 = 3108.95$
Mar. 20	212.94	568.85+212.94=781.79	2	$2 \cdot 568.85 + 212.94 = 1350.64$
Mar. 29	–250	781.79–250=531.79	9	$9 \cdot 781.79 - 250 = 6786.11$
Apr. 1	132.89	531.79+132.89=664.68	3	$3 \cdot 531.79 + 132.89 = 1728.26$
			10	$10 \cdot 664.68 = 6646.80$
Total				19,620.76

Average daily balance $= \dfrac{19{,}620.76}{31} \approx 632.93$

The average daily balance is $632.93.

31. Recording the balances:

Date	Payment or Purchases ($)	Balance Each Day ($)	Days	Unpaid Balance Times Number of Days ($)
Sept. 10		450		
Sept. 12	47.32	450+47.32=497.32	2	$2 \cdot 450 + 47.32 = 947.32$
Sept. 20	157.55	497.32+157.55=654.87	8	$8 \cdot 497.32 + 157.55 = 4136.11$
Sept. 28	–175	654.87–175=479.87	8	$8 \cdot 654.87 - 175 = 5063.96$
Sept. 30	56.73	479.87+56.73=536.60	2	$2 \cdot 479.87 + 56.73 = 1016.47$
			10	$10 \cdot 536.60 = 5366$
Total				16,529.86

Average daily balance $= \dfrac{16{,}529.86}{30} \approx 551.00$

The average daily balance is $551.00.

Finance charge $= 551.00 \cdot \dfrac{0.2125}{365} \cdot 30 \approx 9.62$

The finance charge is $9.62.

32. Recording the balances:

Date	Payment or Purchases ($)	Balance Each Day ($)	Days	Unpaid Balance Times Number of Days ($)
Feb. 3		1023		
Feb. 12	68.54	1023+68.54=1091.54	9	$9\cdot 1023+68.54=9275.54$
Feb. 20	52.33	1091.54+52.33=1143.87	8	$8\cdot 1091.54+52.33=8784.65$
Feb. 22	124.75	1143.87+124.75=1268.62	2	$2\cdot 1143.87+124.75=2412.49$
Feb. 28	–450	1268.62–450=818.62	6	$6\cdot 1268.62-450=7161.72$
			3	$3\cdot 818.62=2455.86$
Total				30,090.26

Average daily balance $=\dfrac{30,090.26}{28}\approx 1074.65$

The average daily balance is $1074.65.

Finance charge $=1074.65\cdot\dfrac{0.196}{365}\cdot 28\approx 16.16$

The finance charge is $16.16.

33. $0.04\cdot 750.25=30.01$
The greater of $30.01 and $25 is $30.01.
The minimum payment is $30.01.

34. Using a graphing calculator,
temporarily skip N, I% = 18.5, PV = 4253.67,
PMT = –450, FV = 0, P/Y and C/Y = 12, END.
Move the cursor to N and press ALPHA SOLVE. N ≈ 10.2951
It will take 11 months to repay the credit card debt.

35. Loan fee $=0.011\cdot 15,000=165$
Amount received = 15,000 – 165 = 14,835
The student receives $14,835.

36. Since this is a subsidized loan, interest does not accrue during the period of time when payments on the loan are not being made.

N = 6(12) = 72, I% = 4.1, PV = 17,000,
temporarily skip PMT, FV = 0, P/Y and C/Y = 12, END.
Move the cursor to PMT and press ALPHA SOLVE.
The monthly payment is $266.74.

37. Since this is a nonsubsidized loan, interest accrues during the period of time when payments on the loan are not being made.
The interest is

$$I = Prt$$
$$= 9000(0.049)4 = 1764$$

Amount owed = 9000 + 1764 = 10,764

N = 6(12) = 72, I% 4.9, PV = 10,764,
temporarily skip PMT, FV = 0, P/Y and C/Y = 12, END.
Move the cursor to PMT and press ALPHA SOLVE.
The monthly payment is $172.85.

38. Temporarily skip N, I% = 5.2, PV = 27,000,
PMT = 325, FV = 0, P/Y and C/Y = 12, END.
Move the cursor to N and press ALPHA SOLVE.
It will take approximately 103 months to repay the loan.

39. Deferred interest $= Prt$
$= 36,000\cdot 0.0595\cdot 2 = 4284$
Amount owed = 36,000 + 4284 = 40,284

N = 5(12) = 60, I% = 5.95, PV = 40,284,
temporarily skip PMT, FV = 0, P/Y and C/Y = 12, END.
Move the cursor to PMT and press ALPHA SOLVE.
The monthly payment is $777.87.

40. Deferred interest $= Prt$
$= 7500\cdot 0.0489\cdot 5 = 1833.75$
Amount owed = 7500 + 1833.75 = 9333.75

N = 7(12) = 84, I% = 4.89, PV = 9333.75,
temporarily skip PMT, FV = 0, P/Y and C/Y = 12, END.
Move the cursor to PMT and press ALPHA SOLVE.
The monthly payment is $131.44.

N = 7(12) = 84, temporarily skip I%,
PV = 7500, PMT = 131.44, FV = 0,
P/Y and C/Y = 12, END.
Move the cursor to I% and press ALPHA SOLVE.
The APR is 11.8%.

41. Purchase price = price – 20%(down payment)
= 28,450 – 0.20(28,450) = 22,760
For the $22,760 loan for 3 years,
N = 3(12) = 36, I% = 3.25, PV = 22,760,
temporarily skip PMT, FV = 0, P/Y and C/Y = 12, END.
Move the cursor to PMT and press ALPHA SOLVE.
The monthly payment is $664.40.

42. a. Sales tax = 55,750(0.0675) = 3763.13

b. Total cost = 55,750 + 3763.13 + 558
= 60,071.13
Down payment =60,071.13(0.10) =6007.11

c. Amount financed
= Cost – down payment
= 60,071.13 – 6007.11 = 54,064.02
For the $54,064.02 loan for 4 years,
N = 4(12) = 48, I% = 4.85, PV =54,064.02, temporarily skip PMT, FV = 0, P/Y and C/Y = 12, END.
Move the cursor to PMT and press ALPHA SOLVE.
The monthly payment is $1241.39.

43. There are 4(12) – 32 = 16 payments remaining.
N = 16, I% = 3.15, temporarily skip PV, PMT = 721.52, FV = 0, P/Y and C/Y = 12, END.
Move the cursor to PV and press ALPHA SOLVE.
The car loan payoff is $11,290.74.

44. N = 5(12) = 60, I% = 4, temporarily skip PV, PMT = –425, FV = 0, P/Y and C/Y = 12, END.
Move the cursor to PV and press ALPHA SOLVE.
The maximum he can finance is $23,077.10.

45. $\frac{1 \text{ gallon}}{33 \text{ miles}} = \frac{x \text{ gallons}}{100 \text{ miles}}$
$x \approx 3.0$
The car has rating of 3.0 gallons per 100 miles.

46. $43,000(0.24) = 10,320$
The depreciation is $10,320.
43,000 – 10,320 = 32,680.
The value after 1 year is $32,680.

47. First find the first year interest paid:
Find the payoff amount after 1 year.
N = 60 – 12 = 48, I% = 4.625, temporarily skip PV, PMT = 748.00
Solving for PV = 32,721.18
Amount of principal paid:
40,000 – 32,721.18 = 7278.82
Total payments: 12(748) = 8976.00
Total interest paid in year 1:
8976.00 – 7278.82 = 1697.18
Or, using a spreadsheet program, the first year interest paid is 1696.99.
Due to rounding, the values are not the same.

Find the total cost:
License and registration:
1% of purchase price
= 47,250(0.01) = 472.50
Sales tax:
7.5% of purchase price
= 47,250(0.075) = 3543.75
Depreciation:
18% of purchase price
= 47,250(0.18) = 8505.00
Fuel cost driving 15,000 miles:
Average cost of gasoline $3.03 per gallon
$\frac{15,000 \text{ miles}}{24 \text{ miles/gallon}} \cdot \left(\frac{\$3.03}{\text{gallon}}\right) = 1893.75$
Year 1 total interest paid: = 1697.18
Insurance: = 950.00
Total = $17,062.18

48. Net cost = 33,950 – 7500 = 26,450
Residual value = 35,400(0.55) = 19,470
$F = (26,450 + 19,470)\frac{6.8}{2400}$
≈ 130.1067
Average monthly finance charge is $130.11.
$D = \frac{26,450 - 19,470}{30}$
≈ 232.67
Average monthly depreciation is $232.67.
$P = D + F = 232.67 + 130.11$
$= 362.78$
The monthly lease payment is $362.78.

49. $\frac{325 + 452.36 + 107 + 2010.23}{6125} \approx 0.4726$
The debt-to-income ratio is approximately 47%.

50. a. Using a graphing calculator,
N = 12(30) = 360, I% = 6.75, PV = 255,800, temporarily skip PMT, FV = 0, P/Y and C/Y = 12.
Solving for PMT = 1659.11
The monthly mortgage payment is $1659.11.

b. $1659.11(360) = 597,279.60$
The payments total $597,279.60.

c. 597,279.60 – 255,800 = 341,479.60
The total interest paid is $341,479.60.

51. a. Using a graphing calculator,
N = 12(30) = 360, I% = 4.875, PV = 329,900, temporarily skip PMT, FV = 0, P/Y and C/Y = 12.
Solving for PMT = 1745.86
The monthly mortgage payment is $1745.86.

b. 1745.86(360) = 628,509.60
The payments total $628,509.60.

628,509.60– 329,900 = 298,609.60
The total interest paid is $298,609.60.

52. $415,000(0.008) = 3320$ for a year.
The monthly PMI $= \frac{3320}{12} \approx 276.67$
The PMI adds $276.67 to the monthly cost.

53. $362,000(0.0225) = 8145$
The discount points are $8145.

54. Difference in cost of points:
$258,790(0.0175) - 258,790(0.01) = 1940.925$

Monthly mortgage payment for 4.25%
N = 12(30) = 360, I% = 4.25, PV = 258,790, temporarily skip PMT, FV = 0, P/Y and C/Y = 12. Solving for PMT = 1273.09
The monthly mortgage payment is $1273.09.

Monthly mortgage payment for 4%
N = 12(30) = 360, I% = 4, PV = 258,790, temporarily skip PMT, FV = 0, P/Y and C/Y = 12. Solving for PMT = 1235.50
The monthly mortgage payment is $1235.50.

The difference in payments is
1273.09 – 1235.50 = 37.59

Months to recoup costs $\dfrac{1940.925}{37.59} \approx 51.634$
It will take about 52 months to recoup the cost.

55. N = 360 – 60 = 300, I% = 3.125, temporarily skip PV, PMT = –2548.83, FV = 0, P/Y and C/Y = 12. Solving for PV = 530,190.71

N = 300, I% = 4.75, PV = 530,190.71, temporarily skip PMT, FV = 0, P/Y and C/Y = 12. Solving for PMT = 3022.71
The new monthly payment will be $3022.71.

56. a. N = 12(25) = 300, I% = 7.5, PV = 189,000, temporarily skip PMT, FV = 0, P/Y and C/Y = 12. Solving for PMT = 1396.69
The mortgage payment is $1396.69.

b. N = 300 – 120 = 180, I% = 7.5, temporarily skip PV, PMT = –1396.69, FV = 0, P/Y and C/Y = 12.
Solving for PV = 150,665.74
The payoff amount is $150,665.74.

57. Fee for points $= 375,000 \cdot 0.0075 = 2812.50$
Loan origination fee $= 375,000 \cdot 0.005 = 1875$
Adjusted loan balance
= 375,000 + 2812.50 + 1875 = $379,687.50

Find the monthly payment.
N = 12(30) = 360, I% = 3.5, PV = 379,687.50, temporarily skip PMT, FV = 0, P/Y and C/Y = 12.
Solving for PMT = 1704.97
The monthly payment is $1704.97.

Change PV from 379,687.50 to 375,000.
N = 300, temporarily skip I%, PV = 375,000, PMT = 1704.97, FV = 0, P/Y and C/Y = 12. Solving for I% = 3.6002
The APR for the loan is 3.600%.

58. Using a graphing calculator,
N = 12(15) = 180, I% = 7, PV = 278,950, temporarily skip PMT, FV = 0, P/Y and C/Y = 12. Solving for PMT = 2507.28
Monthly tax and insurance
$\dfrac{1134+681}{12} = 151.25$
2507.28 + 151.25 = 2658.53
The total monthly payment is $2658.53.

59. Dividend per share = $1.28,
Number of shares = 700
Dividends $= 1.28(700) = 896$
The shareholder earned $896 in dividends.

60. Use the formula $I = Prt$, where $I = 2.31$, $P = 75.61$, and $t = 1$.
$I = Prt$
$2.31 = 75.61r(1)$
$r \approx 0.03055$
The dividend yield is 3.1%.

61. Earnings = $2.2 billion,
number of shares = 500 million
$\text{EPS} = \dfrac{2,200,000,000}{500,000,000} = 4.4$
The EPS for the company is $4.40.

62. EPS = $2.96, stock price = $61.34
$\text{P/E} = \dfrac{\text{Stock price}}{\text{EPS}} = \dfrac{61.34}{2.96} \approx 20.7$
The P/E is 20.7.

63. Interest on 4.5% bond = Interest on 4% bond
$1000(0.045) = x(0.04)$
$45 = 0.04x$
$x = 1125$
The nominal value of the bond is $1125.

64. Net asset value of one share
$= \dfrac{\$5,600,000,000}{62,000,000} \approx 90.32$
The net asset value of one share is $90.32.

65. Fee $= 0.0325 \cdot \$10,000 = \$325$
Funds available = $10,000 – $325 = $9675
Number of shares purchased
$= \dfrac{\$9675}{\$51.86} \approx 186.560$
The manager can purchase 186.560 shares.

66. Real rate of return $= \dfrac{1+\text{nominal rate}}{1+\text{inflation rate}} - 1$
$= \dfrac{1+0.07}{1+0.021} - 1$
≈ 0.047992
The real rate of return is 4.8%.

Chapter 5: Linear Functions

THINK ABOUT IT, SECTION 5.1

1. $z = 7$

2. $x = -2$ and $y = 5$

3. No. That would mean that (2, 3) and (2, 4) belong to the function, which means that the same value of the independent variable, 2, is paired with two different values for the dependent variable. This contradicts the *one and only one* value of the dependent variable for a value of the independent variable.

4. Yes. Different values of the independent variable can be paired with the same value of the dependent variable. This cannot be reversed; that is, the same value of the independent variable cannot be paired with two different values of the dependent variables. See the preceding Exercise 3.

5. The independent variable is *x*. The dependent variable is *y*. The name of the function is *f*.

6. The **domain** of a function is the set of all the first components of the ordered pairs of the function. The **range** of the function is the set of all second components of the ordered pairs of the function.

EXERCISE SET 5.1

1. The independent variable is *h*, the height of the ball before it is dropped. The dependent variable is *s*, the speed of the ball when it hits the ground. The function is $s = sP(h)$.

3. The independent variable is *T*, the temperature of the air. The dependent variable is *s*, the speed of sound. The function is $s = v(T)$.

5. $V(t) = 75{,}000 - 7500t$
   $V(4) = 75{,}000 - 7500(4) = 45{,}000$
   The value of the tractor after 4 years is \$45,000.

7. $f(s) = s^2$
   $f(12) = 12^2 = 144$
   The area of a square with side 12 is 144 cm^2.

9. a. $h(t) = -16t^2 + 80t + 4,\ 0 \le t \le 5.05$
   $h(2) = -16(2)^2 + 80(2) + 4 = 100$
   The height of the ball after 2 s is 100 ft.

   b. $h(t) = -16t^2 + 80t + 4,\ 0 \le t \le 5.05$
   $h(4) = -16(4)^2 + 80(4) + 4 = 68$
   The height of the ball after 4 s is 68 ft.

11. a. $s(t) = \dfrac{1087\sqrt{t+273}}{16.52}$
   $s(0) = \dfrac{1087\sqrt{0+273}}{16.52} \approx 1087.178$
   When the temperature is 0°C, the speed of sound in air is 1087 ft/s.

   b. $s(t) = \dfrac{1087\sqrt{t+273}}{16.52}$
   $s(25) = \dfrac{1087\sqrt{25+273}}{16.52} \approx 1135.867$
   When the temperature is 25°C, the speed of sound in air is 1136 ft/s.

13. a. $T(L) = 2\pi\sqrt{\dfrac{L}{32}}$
   $T(3) = 2\pi\sqrt{\dfrac{3}{32}} \approx 1.9238$
   A pendulum with length 3 ft takes 1.92 s to make one swing.

   b. Recall 9 inches $= \dfrac{9}{12}$ feet $= \dfrac{3}{4}$ feet
   $T(L) = 2\pi\sqrt{\dfrac{L}{32}}$
   $T\left(\dfrac{3}{4}\right) = 2\pi\sqrt{\dfrac{3/4}{32}} \approx 0.9619$
   A pendulum with length 9 inches takes 1.0 s to make one swing.

15. a. The independent variables are *p*, *H*, and *A*.

   b. The dependent variable is *T*.

   c. Here, $p = 0.65$, $A = 31$, $H = 70$
   $T = p(220 - A) + H(1 - p)$
   $T = 0.65(220 - 31) + 70(1 - 0.65)$
   $= 147.35$
   The target heart rate is 147 beats per minute.

17. a. The independent variables are *w*, *h*, and *a*.

   b. The dependent variable is BMR.

   c. Here, $w = 77$, $h = 170$, and $a = 28$
   BMR
   $= 13.4w + 4.8h - 5.7a + 88.4$
   $= 13.4(77) + 4.8(170) - 5.7(28) + 88.4$
   $= 1776.6$
   The BMR is 1777 Calories.

19. a. $t = f(s) = 2.5s$
$60 = 2.5s$
$s = 24$
The ordered pair is (24, 60).

b. It takes 32 seconds to download an 80-gigabyte file.

c. Graphing the function for $8 \le s \le 40$.

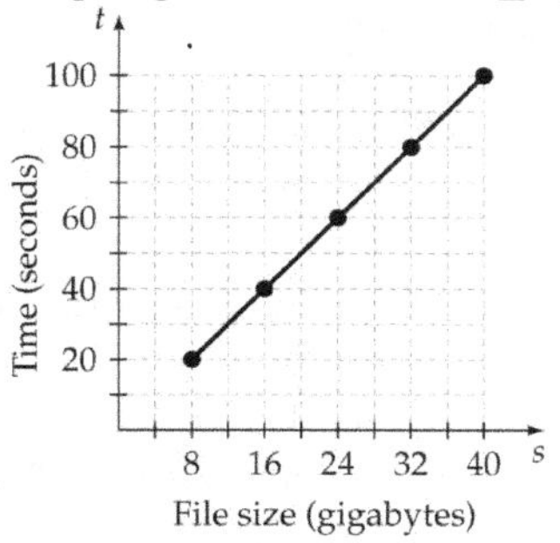

21. a. $N = -0.07p^2 + 0.8p + 90$
$N = -0.07(20)^2 + 0.8(20) + 90 = 78$
The ordered pair is (20, 78).

b. If the price of the game is $25, the company expects to sell 66,000 games.

c. Graphing the function for $10 \le p \le 40$.

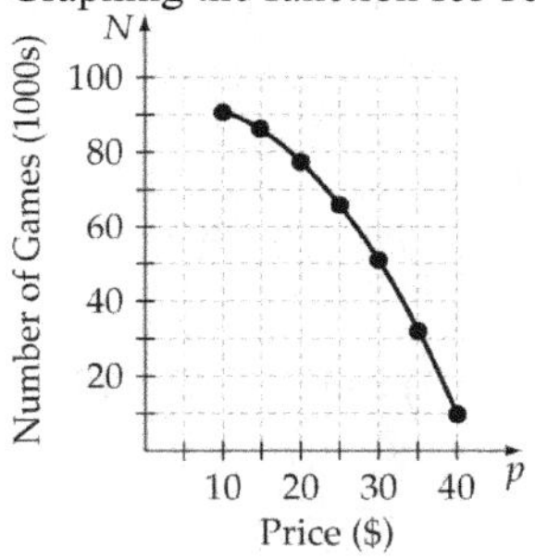

23. Approximately 24 feet

25. The maximum height of the ball is at a little more than 200 yards.

27. a. $\dfrac{700-1200}{3-1} = -250$
The average rate of change is –250 bacteria per hour.

b. The bacteria population is decreasing at an average rate of 250 bacteria per hour between 1 and 3 hours after measurements begin.

29. a. $\dfrac{19,400-16,300}{5-2} \approx 1033.33$
The average rate of change is 1033 flu cases per week.

b. The number of flu cases is increasing at an average rate of 1033 flu cases per week between 2 and 5 weeks after measurements begin.

31. As the values of x increase, the values of y decrease.

33. *A*

35. a. \$1000/year

b. $f(t) = 10,000 + 1000t$
$f(2) = 10,000 + 1000(2) = 12,000$
The value of the investment after 2 years will be \$12,000.

37. a. –2 lb/week

b. $f(t) = -2t + 180$
$f(5) = -2(5) + 180 = 170$
After 7 weeks, the person's weight is 170 pounds.

39. a. For each wind speed, there is *one and only one* F-scale.

b. The domain is 40 mph to 318 mph.

c. For a wind speed of 200 mph, the F-scale value is F3.

THINK ABOUT IT, SECTION 5.2

1. a. $y = \dfrac{1}{x}$ is not a linear function since the variable x is in the denominator.

b. $y = x$ is a linear function.
Here, $m = 1$ and $b = 0$.

c. $F(v) = 4 - 0.256v$ is a linear function.
Here, $m = -0.256$ and $b = 4$.

d. $B(n) = 3n - n^2$ is not a linear function since there is a squared term, n^2.

2. A line with 0 slope is horizontal. A line with no slope is vertical.

3. (0,–3) is the y-intercept.
(2, 0) is the x-intercept.

4. For parallel lines, yes, the slopes are the same. No, the y-intercepts are different.

5. *A* has the better running speed.

6. *B* is depreciating more quickly.

7. b and d are not examples of slope.

EXERCISE SET 5.2

1. For $y = 2x - 3$, $m = 2$ and $b = -3$.

x	$y = 2x - 3$		y	(x, y)
-1	$y = 2(-1) - 3$	$=$	-5	$(-1, -5)$
0	$y = 2(0) - 3$	$=$	-3	$(0, -3)$
3	$y = 2(3) - 3$	$=$	3	$(3, 3)$

Plot the ordered pairs and draw a line through the points.

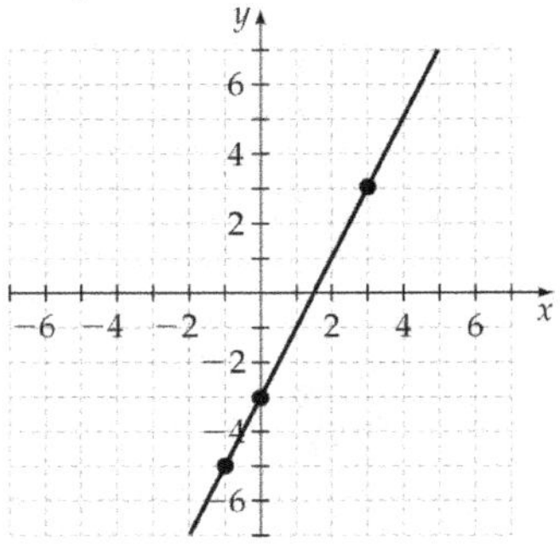

3. For $B(n) = 3 - \frac{n}{4}$, $m = -\frac{1}{4}$ and $b = 3$.

n	$B(n) = 3 - \frac{n}{4}$		$B(n)$	$(n, B(n))$
-4	$3 - \frac{-4}{4}$	$=$	4	$(-4, 4)$
0	$3 - \frac{0}{4}$	$=$	3	$(0, 3)$
4	$3 - \frac{4}{4}$	$=$	2	$(4, 2)$

Plot the ordered pairs and draw a line through the points.

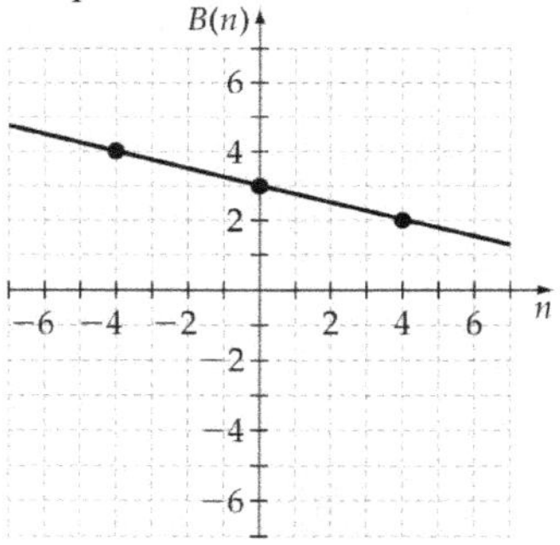

5. $(x_1, y_1) = (-2, 4)$ and $(x_2, y_2) = (4, -1)$

$$m = \frac{y_2 - y_1}{x_2 - x_1} = \frac{-1 - 4}{4 - (-2)} = \frac{-5}{6} = -\frac{5}{6}$$

7. $(x_1, y_1) = (5, 7)$ and $(x_2, y_2) = (-1, 7)$

$$m = \frac{y_2 - y_1}{x_2 - x_1} = \frac{7 - 7}{-1 - 5} = \frac{0}{-6} = 0$$

9. $(x_1, y_1) = (1, 2)$ and $(x_2, y_2) = (2, 5)$

$$m = \frac{y_2 - y_1}{x_2 - x_1} = \frac{5 - 2}{2 - 1} = \frac{3}{1} = 3$$

$(x_1, y_1) = (2, 5)$ and $(x_2, y_2) = (4, 11)$

$$m = \frac{y_2 - y_1}{x_2 - x_1} = \frac{11 - 5}{4 - 2} = \frac{6}{2} = 3$$

Yes. The slope between any two points is the same: 3 gallons/minute.

11. $(x_1, y_1) = (1, 85)$ and $(x_2, y_2) = (2, 133)$

$$m = \frac{y_2 - y_1}{x_2 - x_1} = \frac{133 - 85}{2 - 1} = 48$$

$(x_1, y_1) = (2, 133)$ and $(x_2, y_2) = (3, 149)$

$$m = \frac{y_2 - y_1}{x_2 - x_1} = \frac{149 - 133}{3 - 2} = 16$$

No. The slope between points 1 and 2 does not equal the slope between points 2 and 3.

13. $(x_1, y_1) = (2, 80)$ and $(x_2, y_2) = (6, 240)$

$$m = \frac{y_2 - y_1}{x_2 - x_1} = \frac{240 - 80}{6 - 2} = \frac{160}{4} = 40$$

$m = 40$ mi/hr

The motorist was traveling at 40 mph.

15. $(x_1, y_1) = (10, 20)$ and $(x_2, y_2) = (25, 5)$

$$m = \frac{y_2 - y_1}{x_2 - x_1} = \frac{5 - 20}{25 - 10} = \frac{-15}{15} = -1$$

$m = -1000$ ft/min

The height of the plane is decreasing at a rate of 1000 feet per minute.

17. $(x_1, y_1) = (2, 3.7)$ and $(x_2, y_2) = (5, 1.8)$

$$m = \frac{y_2 - y_1}{x_2 - x_1} = \frac{1.8 - 3.7}{5 - 2} \approx -0.63$$

$m = -0.63$ million gallons/minute

The water in the lock is decreasing by 0.63 million gallons each minute.

19. $(x_1, y_1) = (4, 290)$ and $(x_2, y_2) = (16, 110)$

$$m = \frac{y_2 - y_1}{x_2 - x_1} = \frac{110 - 290}{16 - 4} = -15$$

$m = -15$ °F/min

The temperature inside the oven is decreasing at a rate of 15°F per minute.

21. $(x_1, y_1) = (0, 62)$ and $(x_2, y_2) = (2000, 56)$

$$m = \frac{y_2 - y_1}{x_2 - x_1} = \frac{56 - 62}{2000 - 0} = -0.003$$

$m = -0.003$°F/foot

$$100 \text{ feet} \left(\frac{-0.003°\text{F}}{\text{foot}}\right) = -0.3°\text{F}$$

For every 100 feet, the temperature changes –0.3°F.

23. No. A 1-to-12 ratio is $\frac{1}{12} \approx 0.083$, and $0.09 > 0.083$.

25. a. $(x_1, y_1) = (100,\ 25)$ and $(x_2, y_2) = (400,\ 60)$
$$m = \frac{y_2 - y_1}{x_2 - x_1} = \frac{60 - 25}{400 - 100} \approx 0.11667$$
The evaporation rate is 0.117 gallon/square foot.

b. If an area of 100 square feet has 25 gallons evaporate with a rate of 0.117 gallon per square foot, then an area of 200 square feet (100 square feet more) will have an additional evaporation of
$$100(0.117) = 11.7$$
So a total of 25 + 11.7, or 36.7 gallons of water will evaporate.

27. a. $(x_1, y_1) = (2000,\ 50)$ and $(x_2, y_2) = (3000,\ 40)$
$$m = \frac{y_2 - y_1}{x_2 - x_1} = \frac{40 - 50}{3000 - 2000} = -0.01$$
The slope is –0.01 dollar/drill sold.

b. For each 1-cent decrease in price one additional drill can be sold.

c. If 3000 drills can be sold for $40 each with a rate of decrease of –0.01 dollar per drill sold, then selling drills for $38 each ($2 dollars less) will sell
$$-2 \text{ dollars} \div \left(\frac{-0.01 \text{ dollar}}{\text{drill sold}}\right) = 200 \text{ drills}$$
So a total of 3000 + 200, or 3200 drills will be sold.

29. a. The N-intercept is (0, 270).

b. If the tuition were $0, then 270 students would enroll.

c.
$$N = -0.3t + 270$$
$$0 = -0.3t + 270$$
$$0.3t = 270$$
$$t = 900$$
The t-intercept is (900, 0).

d. If the tuition were $900, no students would enroll.

31. a. The P-intercept is (0,–1000).

b. If the business sells 0 smartwatches, it will lose $1000.

c.
$$P = 25n - 1000$$
$$0 = 25n - 1000$$
$$1000 = 25n$$
$$n = 40$$
The n-intercept is (40, 0).

d. The company must sell 40 smartwatches to break even.

THINK ABOUT IT, SECTION 5.3

1. a. Decrease
   b. Increase
   c. Decrease
   d. Increase

2. °F/min

3. Bucket B

4. In 4 hrs, bucket B will be empty ($y = 0$).

EXERCISE SET 5.3

1. $T = 0.02s + 4000$
When sales are $80,000,
$$T = 0.02(80{,}000) + 4000 = 5600$$
the executive will earn $5600.

3. $d = 2700 - 500t$
After 2 hours,
$$d = 2700 - 500(2) = 1700$$
the plane is 1700 miles away from its destination.

5. $A - 12{,}000 = 1250(t - 8)$
$$A = 1250t + 2000$$
After 20 minutes,
$$A = 1250(20) + 2000 = 27{,}000$$
the altitude is 27,000 feet.

7.
$$S = mh + b$$
$$2250 = -34(68) + b$$
$$b = 4562$$
So the function is $S = -34h + 4562$.
A person 60 inches tall,
$$S = -34(60) + 4562 = 2522$$
will take 2522 steps.

9. $m = \frac{24 - 3}{6 - 0} = 3.5$
$b = 3$
So the function is $H = 3.5n + 3$.
In 2020, which is 10 years after 2010,
$$H = 3.5(10) + 3 = 38$$
the tree will be 38 feet tall.

11. $m=\frac{29-33}{2.5-2}=-8$
$mpg=mw+b$
$33=-8(2)+b$
$b=49$
So the function is $mpg=-8w+49$.
A car that weighs 3 tons
$mpg=-8(3)+49=25$
has a fuel efficiency of 25 mpg.

13. a. $F(t)=99+25t$
After 2 years, which is 24 months,
$F(24)=99+25(24)=699$
the total cost is \$699.

b. Since \$699 < \$749, it is less than.

15. $C(x)=85x+30{,}000$
A house with floor space of 1800 ft^2
$C(1800)=85(1800)+30{,}000=183{,}000$
will cost \$183,000.

17. $m=\frac{55{,}000-50{,}000}{17{,}500-18{,}000}=-10$
$C=mx+b$
$50{,}000=-10(18{,}000)+b$
$b=230{,}000$
So the function is $C(x)=-10x+230{,}000$.
For a price of \$17,000 per car,
$C(17{,}000)=-10(17{,}000)+230{,}000$
$=60{,}000$
60,000 cars will be sold.

19. a. $m=\frac{76-64}{10.0-8.5}=8$
$y=mx+b$
$76=8(10.0)+b$
$b=-4$
So the function is $y=8x-4$.

b. For a reading test grade of 10.5,
$y=8(10.5)-4=80$
will result in a final exam grade of 80.

21. a. $m=\frac{2300-600}{60-10}=34$
$y=mx+b$
$600=34(10)+b$
$b=260$
So the function is $y=34x+260$.

b. After playing for 25 minutes,
$y=34(25)+260=1110$
1110 ml of water will be lost.

23. a. $m=\frac{21-80}{56-22}\approx-1.735$
$y=mx+b$
$21=-1.735(56)+b$
$b=118.16$
So the function is $y=-1.735x+118.16$.

b. At a latitude of 45°N,
$y=-1.735(45)+118.16=40.085$
will result in a temperature of 40°F.

THINK ABOUT IT, SECTION 5.4

1. It is a line that approximates a bivariate data set.
2. No. They have perfect correlation, but the slope of the least-squares line is negative.
3. Decreases
4. Positive

EXERCISE SET 5.4

1. a. Positive linear correlations between the x and y variables are diagrams a and b.

b. Negative linear correlations between the x and y variables are diagrams c and d.

3. a. Enter the data using either a calculator or spreadsheet. Then calculate the least-squares line.
$\hat{y}=-0.1700621321x+54{,}545.58535$

b. Substitute $x=30{,}000$ in the least-squares line equation.
$\hat{y}=-0.1700621321(30{,}000)$
$+54{,}545.58535$
$\approx 49{,}400$
The retail price is \$49,400.

c. $r=-0.9993480085$

d. The variables are negatively correlated, which means that as miles go up, retail value goes down.

5. a. Enter the data using either a calculator or spreadsheet. Then calculate the least-squares line.
$\hat{y}=0.0385003892x+10.66206002$

b. Substitute $x=200$ in the least-squares line equation.
$\hat{y}=0.0385003892(200)+10.66206002$
≈ 18.362
A wave that is 200 m long will have speed 18.4 meters/second.

7. a. Enter the data using either a calculator or spreadsheet. Then calculate the least-squares line.
$\hat{y} = -0.3396749522x + 25.00745698$

b. Substitute $x = 30$ in the least-squares line equation.
$\hat{y} = -0.3396749522(30) + 25.00745698$
≈ 14.8172
A person 30 years old spends about 15 hours on the Internet.

c. $r = -0.9987037488$

d. Yes. $|r| > 0.9$

9. a. Enter the data using either a calculator or spreadsheet. Then calculate the least-squares line.
$\hat{y} = -0.9091666667x + 79.20833333$

b. Substitute $x = 25$ in the least-squares line equation.
$\hat{y} = -0.9091666667(25) + 79.20833333$
≈ 56.479
A woman of age 25 has approximately 56 years remaining.

c. Interpolation

11. a. Enter the data using either a calculator or spreadsheet. Then calculate the least-squares line.
$\hat{y} = 8.73333333x + 60.6444444$

b. Substitute $x = 5$ in the least-squares line equation.
$\hat{y} = 8.73333333(5) + 60.6444444$
$= 104.311111$
Someone traveling 5 mph, the pulse is approximately 104 bpm.

c. $r = 0.9996408517$, so $|r| > 0.9$.
Yes there is a strong correlation between speed and bpm.

CHAPTER 5 REVIEW EXERCISES

1. The independent variable is s, the size of the file. The dependent variable is t, the time to upload a file.
The function is $t = f(s)$.

2. $N(p) = 10{,}000 - 5p$
$N(1000) = 10{,}000 - 5(1000) = 5000$
The company can expect to sell 5000 cameras.

3. $h(t) = -16t^2 + 70t + 5$
$h(3) = -16(3)^2 + 70(3) + 5 = 71$
The height of the ball after 3 s is 71 ft.

4. $T(m) = 70 + 30 \cdot 0.7^m$
$T(2) = 70 + 30 \cdot 0.7^2 = 84.7$
After 5 minutes, the temperature of the soup is approximately 85°F.

5. a. $L = f(x) = 3 + 0.25x$
$L = f(8) = 3 + 0.25(8) = 5$
The ordered pair is (8, 5).

b. When a weight of 12 ounces is attached to the spring, its length is 6 inches.

c. Graphing the function for $0 \le x \le 12$.

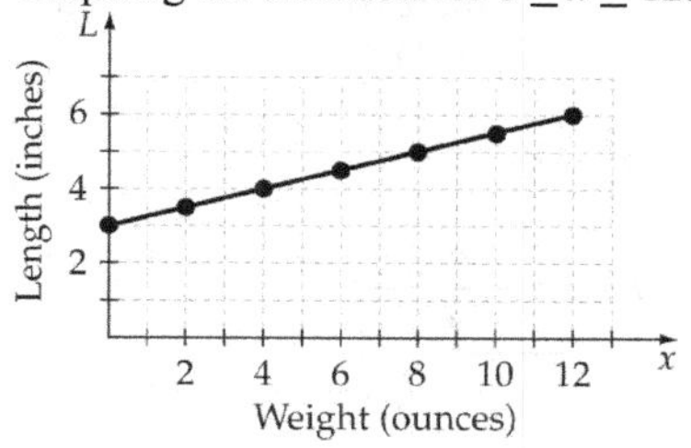

6. a. $P = g(x) = -0.01x^2 + 4x - 100$
$P = g(200) = -0.01(200)^2 + 4(200) - 100$
$= 300$
The ordered pair is (200, 300).

b. When the company produces 250,000 parts, the expected profit is $275,000.

c. Graphing the function for $50 \le x \le 350$. Recall that x is in thousands.

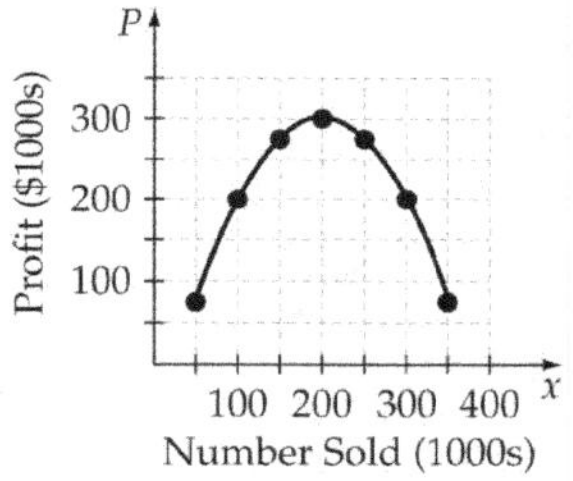

7. From the graph, the amount borrowed is approximately $27,000.

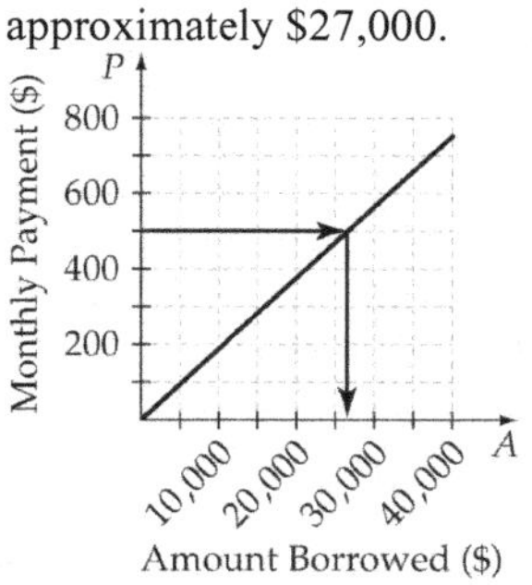

8. a. $\dfrac{25-44}{90-70} = -0.95$
The average rate of change is −0.95 ft/sec per foot.

b. The velocity of the ball is decreasing at an average rate of 0.95 foot per second for each 1 foot of additional height.

9. The velocity is decreasing the fastest between points B_1 and B_2.

10. a. $\frac{2950-2320}{14-10}=157.5$

The average rate of change is approximately 158 people/day.

b. The number of people who have heard the rumor is increasing at an average rate of 158 people/day.

11. Greater than

12. For $f(x)=-\frac{3}{4}x+1$

x	$f(x)=-\frac{3}{4}x+1$		$f(x)$	$(x, f(x))$
−4	$-\frac{3}{4}(-4)+1$	=	4	(−4, 4)
0	$-\frac{3}{4}(0)+1$	=	1	(0, 1)
4	$-\frac{3}{4}(4)+1$	=	−2	(4, −2)

Plot the ordered pairs and draw a line through the points.

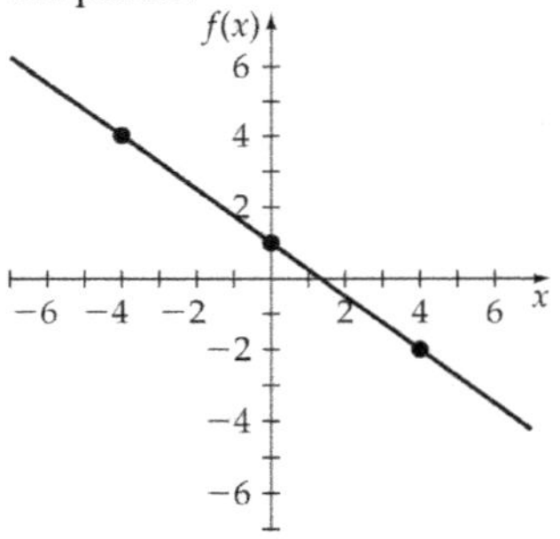

13. For $z=\frac{2}{3}t-3$

t	$z=\frac{2}{3}t-3$		z	(t, z)
−3	$\frac{2}{3}(-3)-3$	=	−5	(−3, −5)
0	$\frac{2}{3}(0)-3$	=	−3	(0, −3)
3	$\frac{2}{3}(3)-3$	=	−1	(3, −1)

Plot the ordered pairs and draw a line through the points.

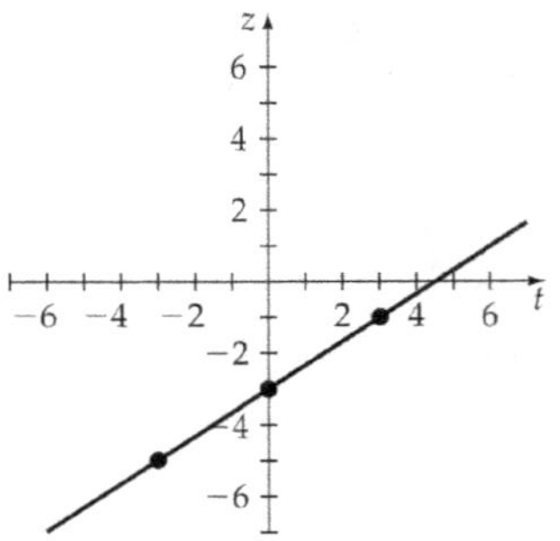

14. $(x_1, y_1)=(3,\ 2)$ and $(x_2, y_2)=(2, -3)$

$m=\frac{y_2-y_1}{x_2-x_1}=\frac{-3-2}{2-3}=\frac{-5}{-1}=5$

15. $(x_1, y_1)=(-1,\ 4)$ and $(x_2, y_2)=(-3, -1)$

$m=\frac{y_2-y_1}{x_2-x_1}=\frac{-1-4}{-3-(-1)}=\frac{-5}{-2}=\frac{5}{2}$

16. $(x_1, y_1)=(2, -5)$ and $(x_2, y_2)=(-4, -5)$

$m=\frac{y_2-y_1}{x_2-x_1}=\frac{-5-(-5)}{-4-2}=\frac{0}{-6}=0$

17. $(x_1, y_1)=(5,\ 2)$ and $(x_2, y_2)=(5,\ 7)$

$m=\frac{y_2-y_1}{x_2-x_1}=\frac{7-2}{5-5}=\frac{5}{0}$ undefined; no slope

18. $(x_1, y_1)=(40,\ 13)$ and $(x_2, y_2)=(180,\ 6)$

$m=\frac{y_2-y_1}{x_2-x_1}=\frac{6-13}{180-40}=-0.05$

The car is using gas at a rate of 0.05 gallon per mile driven.

19. a. $(x_1, y_1)=(400,\ 800)$ and $(x_2, y_2)=(600,\ 1800)$

$m=\frac{y_2-y_1}{x_2-x_1}=\frac{1800-800}{600-400}=5$

b. Profit is increasing at a rate of \$5 per rose bush sold.

c. Difference in rose bushes sold:
350 – 300 = 50
50(5) + 300 = 550
The profit to sell 350 rose bushes is \$550.

20. $\frac{40-0}{0-10}=-4$

The average rate of change per second is –4 feet/second.

21. The vertical axis intercept is (0, 25,000). The original price of the van was \$25,000.
The horizontal axis intercept is (5, 0). The depreciated value of the van after 5 years is \$0.

22. a. $C = 12x + 100$

    b. $C = 12x + 100$
    $C = 12(288) + 100 = 3556$
    To carpet 288 square feet the cost is \$3556.

23. $y = mt + b$
    $200 = 15(10) + b$
    $b = 50$
    So the function is $y = 15t + 50$.
    After 30 minutes,
    $y = 15(30) + 50 = 500$
    500 milliliters of hydrogen peroxide will be produced.

24. a. $m = \dfrac{-34 - 2}{8 - 2} = -6$
    $y = mx + b$
    $2 = -6(2) + b$
    $b = 14$
    So the function is $y = -6x + 14$.

    b. At an altitude of 7 km,
    $y = -6(7) + 14 = -28$
    will result in a temperature of –28°C.

25. a. $m = \dfrac{93 - 16}{60 - 10} = 1.54$
    $y = mt + b$
    $16 = 1.54(10) + b$
    $b = 0.6$
    So the function is $y = 1.54t + 0.6$.

    b. After 35 minutes of a high-intensity workout,
    $y = 1.54(35) + 0.6 = 54.5$
    uses 54.5 grams of carbohydrates.

26. Enter the data using either a calculator or spreadsheet. Then calculate the least-squares line for parts a and b.
    a. $r = 0.9992342081$

    b. $\hat{y} = 0.0725547445x + 0.2883211679$

    c. Substitute $x = 195$ in the least-squares line equation.
    $\hat{y} = 0.0725547445(195) + 0.2883211679$
    ≈ 14.4
    The 195 lb weight will stretch the spring 14.4 inches.

Chapter 6: Nonlinear Functions

Think About It, Section 6.1

1. Since the base of the exponent is 1.03, which is greater than 1, this is exponential growth. The percentage of growth is 3%.

2. Since the base of the exponent is 0.75, which is less than 1, this is exponential decay. The percentage of decay is 1 – 0.75 = 0.25, or 25%.

3. False. After 5 years,
at 2% the amount of interest is \$10.41.
at 4% the amount of interest is \$21.67.
The amount of interest at 4% is more than twice the amount of interest at 2%.

4. 50 seconds

5. 5.5 hours

Exercise Set 6.1

1. $A_0 = 2500,\ r = \dfrac{0.075}{12} = 0.00625,\ T = 3t$, and
$t = 12$, so $T = 3(12) = 36$
$$\begin{aligned} A &= A_0(1+r)^T \\ &= 2500(1+0.00625)^{36} \\ &\approx 3128.62 \end{aligned}$$
The value of the investment after 3 years is \$3128.62.

3. $A_0 = 3{,}000{,}000,\ r = 0.015,\ T = 10$
$$\begin{aligned} A &= A_0(1+r)^T \\ &= 3{,}000{,}000(1+0.015)^{10} \\ &\approx 3{,}500{,}000 \text{ people} \end{aligned}$$
The population will be 3,500,000 people after 10 years.

5. $A = 2(3)^{t/8}$

7. $r = 1 - 0.19 = 0.81$
$P = 100(0.81)^d$
At a depth of 7 meters,
$P = 100(0.81)^7 \approx 22.9$
the percentage of light is 22.9%.

9. $r = 1 - 0.36 = 0.64$
$C = 300{,}000(0.64)^{t/2}$
For 1 GB of storage in 2018,
$t = 2018 - 1980 = 38$
$C = 300{,}000(0.64)^{38/2} \approx 62$
the cost is \$62.

11. a. $r = 1 + 0.20 = 1.2$
$N = 127(1.2)^{n/2}$

b. After 5 days (120 hours)
$N = 127(1.2)^{120/2} \approx 7{,}156{,}134$
there were approximately 7.2 million viewers. So the video would indeed be considered to have gone viral.

12. $r = 1 + 0.16 = 1.16$
$N = 1000(1.16)^{t/3}$
After 20 hours,
$N = 1000(1.16)^{20/3} \approx 2690$
there will be about 2690 viruses.

13. $r = 1 - 0.25 = 0.75$
$P = 100(0.75)^{x/4}$
For a vest 100 mm thick,
$P = 100(0.75)^{100/4} \approx 0.075$
about 7.5% of the x-rays will penetrate the vest.

15. $A_0 = 5000,\ r = \dfrac{0.0375}{365},\ T = 365t$, and
$t = 18$, so $T = 365(18) = 6570$
$$\begin{aligned} A &= A_0(1+r)^T \\ &= 5000\left(1+\frac{0.0375}{365}\right)^{6570} \\ &\approx 9819.82 \end{aligned}$$
The value of the investment after 18 years is \$9819.82.

17. $A_0 = 1000,\ r = \dfrac{0.03}{4} = 0.0075,\ T = 4t$, and
for Jan 2018 to Dec 2020, $t = 3$, so $T = 4(3) = 12$
for Jan 2019 to Dec 2020, $t = 2$, so $T = 4(2) = 8$
$$\begin{aligned} A &= A_0(1+r)^T + A_0(1+r)^T \\ &= 1000(1+0.0075)^{12} + 1000(1+0.0075)^8 \\ &\approx 2155.41 \end{aligned}$$
The value of the investment is \$2155.41.

19. $N = 2(3)^n$

21. Find the exponential model. Note: filtering 68%, allows 100 – 68 = 32% to pass through.
$P = 100(0.32)^{x/10}$
For a 25-mm-thick lens, the percent is
$P = 100(0.32)^{25/10} \approx 5.793$
Yes. The percent of light passing through is approximately 6%.

23. $A = A_0(0.5)^{t/h}$
$A = 5(0.5)^{10{,}000/5730} \approx 1.49146$
There will be approximately 1.5 mg in the bone fragment.

25. a. Let $t = 10$ and evaluate.

$$T = 70 + 280(0.8)^{10} \approx 100.065$$

After 10 minutes, the temperature is approximately 100°F.

b. Evaluate the function for $t = 20$, 30, and 40.

$$T = 70 + 280(0.8)^{20} \approx 73.228$$
$$T = 70 + 280(0.8)^{30} \approx 70.347$$
$$T = 70 + 280(0.8)^{40} \approx 70.037$$

No. The temperature of the oven cannot go below room temperature.

27. Enter the data into a graphing calculator. In the first column, L1, put value of t, where t is the years after 2014, and the values of NO_2 in the second column, L2.
The exponential regression equation is

$$C = 40.8114(0.9750)^t$$

Substitute $t = 2025 - 2014 = 11$ in the equation.

$$C = 40.8114(0.9750)^{11}$$
$$C \approx 30.8910$$

The concentration in 2025 is approximately 31 ppb.

29. Enter the data into a graphing calculator. The values of Altitude in the first column, L1, and the values of Pressure in the second column, L2.
The exponential regression equation is

$$P = 14.8384(0.9775)^h$$

Substitute $h = 15$ in the equation.

$$P = 14.8384(0.9775)^{15}$$
$$P \approx 10.5$$

At an altitude of 15 km, the pressure is approximately 10.5 pounds per square inch.

31. In exercise 29, the function is $P = 14.8384(0.9775)^h$, which is an example of exponential decay. The percentage of decrease is $1 - 0.9775 = 0.0225$, or 2.25%.

THINK ABOUT IT, SECTION 6.2

1. $x = \log(7)$

2. $10^y = 9$

3. Earthquake A having a magnitude of 2 on the Richter scale has an intensity of $10^2 = 100$. Earthquake B having a magnitude of 4 on the Richter scale has an intensity of $10^4 = 10,000$. No, the intensity of B is not twice the intensity of A. The intensity of B is 100 times the intensity of A.

4. Acidic

5. The doubling time is the time it takes for the investment to double in value.

EXERCISE SET 6.2

1. a. $\log x = 5$

$$10^5 = x$$
$$x = 100,000$$

b. $\log x = -1$

$$10^{-1} = x$$
$$x = 0.1$$

c. $x = \log 10,000 = 4$

d. $x = \log\left(\frac{1}{100}\right) = -2$

3.
$$\text{dB}(I) = 10\log\left(\frac{I}{I_0}\right)$$
$$\text{dB}(320I_0) = 10\log\left(\frac{320I_0}{I_0}\right) = 10\log(320) \approx 25.051$$

The decibel level of the cat is approximately 25 dB.

5.
$$N \approx 1.2589^p$$
$$N \approx 1.2589^{100} \approx 9,979,834,850 \approx 1 \times 10^{10}$$

The average car stereo is about 10^{10} times greater than the threshold level.

7.
$$M = \log I$$
$$M = \log(630,000,000) \approx 8.7993$$

The Richter magnitude was approximately 8.8.

9. Use the half-life equation and its solution.

$$A = A_0\left(\frac{1}{2}\right)^{t/h}$$
$$3.18 = 4\left(\frac{1}{2}\right)^{5/h}$$
$$h = 5 \cdot \frac{\log\left(\frac{1}{2}\right)}{\log\left(\frac{3.18}{4}\right)} \approx 15.107$$

The half-life is approximately 15.1 hours.

11. Use the doubling time equation and its solution.

$$A = A_0(2)^{t/h}$$
$$710{,}000 = 500{,}000(2)^{10/h}$$
$$h = 10 \cdot \frac{\log(2)}{\log\left(\frac{710{,}000}{500{,}000}\right)} \approx 19.767$$

The doubling time is approximately 20 minutes.

13. $$2A_0 = A_0\left(1+\frac{0.02}{365}\right)^{365t}$$
$$2 = \left(1+\frac{0.02}{365}\right)^{365t}$$
$$t = \frac{\log(2)}{365\log\left(1+\frac{0.02}{365}\right)} \approx 34.658$$

The doubling time is approximately 34.7 years.

15. Find the intensity of the first speaker.

$$\text{dB} = 10\log\left(\frac{I}{I_0}\right)$$
$$70 = 10\log\left(\frac{I}{I_0}\right)$$
$$7 = \log\left(\frac{I}{I_0}\right)$$
$$\frac{I}{I_0} = 10^7$$
$$I = 10^7 I_0$$

Three speakers triples the intensity to $3 \cdot 10^7 I_0$. Now find the decibel level of the new intensity.

$$\text{dB} = 10\log\left(\frac{3\cdot 10^7 I_0}{I_0}\right)$$
$$= 10\log(3\cdot 10^7)$$
$$\approx 74.771$$

The decibel level is approximately 75 dB.

17. This is exponential decay and $r = 0.10$. Therefore $p = 1 - 0.10 = 0.90$. Since the rate is every 2 hours, $t = 3/2 = 1.5$.

$$A = A_0(p)^t$$
$$A = 400{,}000(0.9)^{1.5}$$
$$\approx 341{,}525.987$$

After 3 hours there is approximately 341,526 gallons of oil remaining.

19. $I_1 = 10^{7.2}$ $\quad I_2 = 10^{3.7}$

$$\frac{I_1}{I_2} = \frac{10^{7.2}}{10^{3.7}}$$
$$= 10^{7.2-3.7} = 10^{3.5}$$
$$\approx 3162.278$$

The original earthquake was approximately 3162 times as intense as the aftershock.

21. a. $$P = 100\left(\frac{1}{2}\right)^{t/50{,}000}$$
$$P = 100\left(\frac{1}{2}\right)^{20{,}000/50{,}000} \approx 75.8$$

The percent chance that the LED will fail after 20,000 hours is 75.8%.

b. Solve for t.

$$30 = 100\left(\frac{1}{2}\right)^{t/50{,}000}$$
$$t = 50{,}000 \cdot \frac{\log\left(\frac{30}{100}\right)}{\log\left(\frac{1}{2}\right)} \approx 86{,}848$$

The chance of an LED light failing will reach 30% after approximately 87,000 hours.

23. Use the half-life equation and its solution.

$$A = A_0\left(\frac{1}{2}\right)^{t/h}$$
$$7 = 8\left(\frac{1}{2}\right)^{1/h}$$
$$h = 1 \cdot \frac{\log\left(\frac{1}{2}\right)}{\log\left(\frac{7}{8}\right)} \approx 5.191$$

The half-life is approximately 5.2 years.

25. Use the half-life equation and its solution.

$$A = A_0\left(\frac{1}{2}\right)^{t/h}$$
$$3 = 5\left(\frac{1}{2}\right)^{2/h}$$
$$h = 2 \cdot \frac{\log\left(\frac{1}{2}\right)}{\log\left(\frac{3}{5}\right)} \approx 2.714$$

The half-life is approximately 2.7 days.

27. $$2A_0 = A_0(1+0.02)^t$$
$$2 = (1.02)^t$$
$$t = \frac{\log(2)}{\log(1.02)} \approx 35.003$$

The doubling time is approximately 35 years.

29. This is exponential decay and $r = 0.20$. Therefore $p = 1 - 0.20 = 0.80$. The exponential equation is

$$A = 1000(0.80)^{t/5}$$

To get 50 pounds of raw sugar, solve for t.

$$A = 1000(0.80)^{t/5}$$
$$50 = 1000(0.80)^{t/5}$$
$$t = 5 \cdot \frac{\log\left(\frac{50}{1000}\right)}{\log(0.80)} \approx 67.126$$

It will take approximately 67.1 hours.

THINK ABOUT IT, SECTION 6.3

1. Since $a = -0.5 < 1$, the graph of the parabola opens down.

2. Yes

3. a. $y = x^2$ is a quadratic function.

   b. $z = 2^x$ is an exponential function since the variable x is in the exponent.

4. The maximum or minimum of a quadratic function occurs at the **vertex** of the graph of the function.

EXERCISE SET 6.3

1. Since $a = 1 > 0$, the graph opens up.
$$x = -\frac{b}{2a} = -\frac{-10}{2(1)} = 5$$
$$f(5) = (5)^2 - 10(5) = -25$$
The vertex is (5,–25).

3. Since $a = 1 > 0$, the graph opens up.
$$x = -\frac{b}{2a} = -\frac{0}{2(1)} = 0$$
$$f(0) = (0)^2 - 10 = -10$$
The vertex is (0,–10).

5. Since $a = -1 < 0$, the graph opens down.
$$x = -\frac{b}{2a} = -\frac{6}{2(-1)} = 3$$
$$f(3) = -(3)^2 + 6(3) + 1 = 10$$
The vertex is (3, 10).

7. Since $a = 2 > 0$, the graph opens up.
$$x = -\frac{b}{2a} = -\frac{-3}{2(2)} = \frac{3}{4} = 0.75$$
$$f\left(\frac{3}{4}\right) = 2\left(\frac{3}{4}\right)^2 - 3\left(\frac{3}{4}\right) + 7 = \frac{47}{8} = 5.875$$
The vertex is $\left(\frac{3}{4}, \frac{47}{8}\right)$ or (0.75, 5.875).

9. Since $a = -4 < 0$, the graph opens down.
$$x = -\frac{b}{2a} = -\frac{1}{2(-4)} = \frac{1}{8} = 0.125$$
$$f\left(\frac{1}{8}\right) = -4\left(\frac{1}{8}\right)^2 + \left(\frac{1}{8}\right) + 1 = \frac{17}{16} = 1.0625$$
The vertex is $\left(\frac{1}{8}, \frac{17}{16}\right)$ or (0.125, 1.0625).

11. Find the minimum or maximum.
$$x = -\frac{b}{2a} = -\frac{8}{2(1)} = -4$$
$$f(-4) = (-4)^2 + 8(-4) = -16$$
Since $a = 1 > 0$, the vertex is a minimum.
The minimum value is –16.

13. Find the minimum or maximum.
$$x = -\frac{b}{2a} = -\frac{0}{2(3)} = 0$$
$$f(0) = 3(0)^2 = 0$$
Since $a = 3 > 0$, the vertex is a minimum.
The minimum value is 0.

15. Find the minimum or maximum.
$$x = -\frac{b}{2a} = -\frac{8}{2(-1)} = 4$$
$$f(4) = -(4)^2 + 8(4) - 1 = 15$$
Since $a = -1 < 0$, the vertex is a maximum.
The maximum value is 15.

17. Find the minimum or maximum.
$$t = -\frac{b}{2a} = -\frac{-8}{2(2)} = 2$$
$$s(2) = 2(2)^2 - 8(2) - 1 = -9$$
Since $a = 2 > 0$, the vertex is a minimum.
The minimum value is –9.

19. Find the minimum or maximum.
$$n = -\frac{b}{2a} = -\frac{10}{2(-2)} = \frac{5}{2} = 2.5$$
$$m\left(\frac{5}{2}\right) = -2\left(\frac{5}{2}\right)^2 + 10\left(\frac{5}{2}\right) + 4 = \frac{33}{2} = 16.5$$
Since $a = -2 < 0$, the vertex is a maximum.
The maximum value is $\frac{33}{2}$ or 16.5.

21. $A(t) = -4.9t^2 + 90t + 9000$
Microgravity begins and ends at a height of 9000 m.
$$9000 = -4.9t^2 + 90t + 9000$$
$$4.9t^2 - 90t = 0$$
$$t(4.9t - 90) = 0$$
$$t = 0 \quad 4.9t - 90 = 0$$
$$t = 0 \qquad t = \frac{90}{4.9} \approx 18.4$$
The time of microgravity is 18.4 seconds.

23. Let x = length of short side and y = the length of the long side.
Perimeter = $2x + 2y = 100$.
Solve for y.

$$2x+2y=100$$
$$2y=100-2x$$
$$y=50-x$$
$$A=x\cdot y$$
$$A=x(50-x)=50x-x^2$$
$$x=-\frac{b}{2a}=-\frac{50}{2(-1)}=25$$

$x = 25$ feet

$$y=50-x=50-25=25$$

$y = 25$ feet
To maximize area, the garden should be 25 feet by 25 feet.

25. a. $3x+2y=2100$

b. $$3x+2y=2100$$
$$2y=2100-3x$$
$$y=1050-\frac{3}{2}x$$

c. $$A=x\cdot y$$
$$A=x\left(1050-\frac{3}{2}x\right)$$
$$A=1050x-\frac{3}{2}x^2$$

d. $$x=-\frac{b}{2a}=-\frac{1050}{2\left(-\frac{3}{2}\right)}=350$$

$x = 350$ feet

$$y=1050-\frac{3}{2}x$$
$$y=1050-\frac{3}{2}(350)=525$$

$y = 525$ feet

e. $A=x\cdot y=350(525)=183{,}750$
The maximum area is 183,750 square feet.

27. a. $T(t)=-0.7t^2+9.4t+59.3$

$$t=-\frac{b}{2a}=-\frac{9.4}{2(-0.7)}\approx 7 \text{ hours}$$

The temperature is a maximum about 7 hours after 6:00 A.M.
Thus the temperature is a maximum at 1:00 P.M.

b. $T(7)=-0.7(7)^2+9.4(7)+59.3\approx 91$
The maximum temperature is approximately 91°F.

29. $h(t)=-16t^2+128t$

a. $t=-\frac{b}{2a}=-\frac{128}{2(-16)}=4$ seconds

b. $h(4)=-16(4)^2+128(4)=256$ feet

c. $$0=-16t^2+128t$$
$$0=-16t(t-8)$$
$$-16t=0 \quad \text{or} \quad t-8=0$$
$$t=0 \qquad\qquad t=8$$

The projectile hits the ground at $t = 8$ seconds.

31. Let x = the number of televisions sold.

a. $$R(x)=xp=x(6000-45x)$$
$$=6000x-45x^2$$

b. $$P(x)=R(x)-C(x)$$
$$=(6000x-45x^2)-(1500x+3000)$$
$$=-45x^2+4500x-3000$$

c. $$x=-\frac{b}{2a}=-\frac{4500}{2(-45)}=50$$

Thus, 50 televisions should be sold to maximize profit.

d. $$P(50)=-45(50)^2+4500(50)-3000$$
$$=\$109{,}500$$

The maximum profit is \$109,500.

CHAPTER 6 REVIEW EXERCISES

1. Since $1.025 > 1$, the equation represents exponential growth.

2. Since $0.9 < 1$, the equation represents exponential decay.

3. The function is a quadratic function, so it is neither exponential growth nor exponential decay.

4. Since $2 > 1$, the equation represents exponential growth.

5. a. $$f=27.5\cdot 4^{(n-1)/24}$$
$$f(40)=27.5\cdot 4^{(40-1)/24}\approx 262$$

The frequency is 262 vibrations/second.

b. $$\frac{27.5\cdot 4^{(52-1)/24}}{27.5\cdot 4^{(40-1)/24}}=4^{51/24-39/24}$$
$$=4^{12/24}$$
$$=4^{1/2}$$
$$=\sqrt{4}$$
$$=2$$

6. $E = 25(0.997)^{t/24}$

$E(480) = 25(0.997)^{480/24} \approx 23.5$

Twenty days after it is activated, the power supply will produce approximately 23.5 watts.

7. Find the exponential function.

$$A = 50,000\left(\frac{60,000}{50,000}\right)^t$$

$$A = 50,000(1.2)^t$$

Let $t = 3.5$ and evaluate.

$$A = 50,000(1.2)^{3.5} \approx 94,646.458$$

After 3.5 hours, there are approximately 94,646 bacteria.

8. Find the exponential function.
$r = 1 + 0.05 = 1.05$

$$H = 50(1.05)^t$$

After 12 months, the child's height

$$H = 50(1.05)^{12} \approx 89.793$$

is approximately 90 centimeters.

9. Find the exponential function.

$$W = 7.2(2)^{t/15}$$

Let $t = 3$ years $= 36$ months and evaluate.

$$W = 7.2(2)^{36/15} \approx 38.002$$

After 3 years, the child weight is approximately 38 pounds.

10. $A_0 = 5000,\ r = \dfrac{0.05}{12},\ T = 12t.$

$$A = A_0(1+r)^T$$

$$A = 5000\left(1+\frac{0.05}{12}\right)^{12t}$$

After 7 years, $t = 7$, so $T = 12(7) = 84$

$$A = 5000\left(1+\frac{0.05}{12}\right)^{84}$$

$$\approx 7090.18$$

the value of the investment is \$7090.18.

11. $A_0 = 10,000,\ r = \dfrac{0.0375}{365},\ T = 365t.$

$$A = A_0(1+r)^T$$

$$A = 10,000\left(1+\frac{0.0375}{365}\right)^{365t}$$

$$A = 10,000(1.00010274)^{365t}$$

After 8 years, $t = 8$, so $T = 365(8) = 2920$

$$A = 10,000(1.00010274)^{2920}$$

$$\approx 13,498.39$$

the value of the investment is \$13,498.39.

12. a. Find the exponential function.
$r = 1 + 0.12 = 1.12$

$$A = 2000(1.12)^{t/4}$$

b. Let $t = 40$ hours and evaluate.

$$A = 2000(1.12)^{40/4} \approx 6211.696$$

After 40 hours, approximately 6212 people have received the tweet.

13. This is exponential decay and $r = 0.30$.
Therefore $p = 1 - 0.30 = 0.70$.
Since the rate is every 30 minutes, we use $t/30$.
The exponential decay model is

$$A = A_0(p)^t$$

$$A = 25(0.7)^{t/30}$$

When number of pounds of salt after 2 hours, then $t = 2$ hours $= 120$ minutes.

$$A = 25(0.7)^{120/30}$$

$$\approx 6.0025$$

After 2 hours there is approximately 6.00 pounds of salt.

14. This is exponential decay and $r = 0.05$.
Therefore $p = 1 - 0.05 = 0.95$.
Since the rate is every 5 seconds, we use $t/5$.
The exponential decay model is

$$A = A_0(p)^t$$

$$A = 500,000(0.95)^{t/5}$$

Let $t = 60$.

$$A = 500,000(0.95)^{60/5}$$

$$\approx 270,180$$

After 60 seconds there are approximately 270,000 pathogens.

15. Find the exponential function.

$$H = 10(0.90)^n$$

16. Use the half-life equation to find the exponential function.

$$A = A_0(0.5)^{t/h}$$

$$A = 75(0.5)^{t/28.8}$$

Let $t = 50$ and evaluate.

$$A = 75(0.5)^{50/28.8} \approx 22.513$$

After 50 years, approximately 22.5 grams remain.

17. Use the half-life equation to find the exponential function.

$$A = A_0(0.5)^{t/h}$$

$$A = 50(0.5)^{t/15}$$

Let $t = 6$ months $= 0.5$ years and evaluate.

$$A = 50(0.5)^{0.5/15} \approx 48.858$$

After 6 months, approximately 48.9 grams remain.

18. Enter the data into a graphing calculator. The values of Time in the first column, L1, and the values of Bacteria in the second column, L2. The exponential regression equation is

$$A = 495.0498469(1.014927141)^t$$

Substitute $t = 45$ in the equation.

$$A = 495.0498469(1.014927141)^{45}$$
$$A \approx 964.313$$

After 45 minutes there are approximately 964 bacteria.

19. Enter the data into a graphing calculator. The values of Time in the first column, L1, and the values of Value in the second column, L2. The exponential regression equation is

$$A = 35{,}094.45404(0.9729733113)^m$$

Substitute $m = 4$ years (48 months) in the equation.

$$A = 35{,}094.45404(0.9729733113)^{48}$$
$$A \approx 9420.70$$

After 4 years, the car's value is approximately \$9420.70.

20. a. $\log x = 4$

$$10^4 = x$$
$$x = 10{,}000$$

b. $\log x = -3$

$$10^{-3} = x$$
$$x = 0.001$$

c. $x = \log(10{,}000{,}000) = 7$

d. $x = \log\left(\frac{1}{10{,}000}\right) = -4$

21. a. $\log x = \frac{4}{5}$

$$10^{4/5} = x$$
$$x \approx 6.3096$$

b. $\log x = -1.2$

$$10^{-1.2} = x$$
$$x \approx 0.0631$$

c. $x = \log 23 \approx 1.3617$

d. $\log x = 0.15$

$$10^{0.15} = x$$
$$x \approx 1.4125$$

22. $\text{pH} = -\log[\text{H}^+]$

$$\text{pH} = -\log 0.0000002$$
$$\approx 6.69897$$

The pH is 6.7. Because 6.7 < 7, the maple syrup is an acid.

23.

$$10^{-\text{pH}} = [\text{H}^+]$$
$$10^{-5.2} = [\text{H}^+]$$
$$0.00000631 \approx [\text{H}^+]$$

The hydrogen ion concentration is approximately 0.00000631 mole per liter.

24. $M = \log I$

$$M = \log(800{,}000)$$
$$\approx 5.90309$$

The Richter magnitude is approximately 5.9.

25. $I_1 = 10^{7.7} \quad I_2 = 10^{4.3}$

$$\frac{I_1}{I_2} = \frac{10^{7.7}}{10^{4.3}}$$
$$= 10^{7.7-4.3} = 10^{3.4}$$
$$\approx 2511.886$$

The first earthquake was approximately 2512 times as intense as the second earthquake.

26.

$$\text{dB}(I) = 10\log\left(\frac{I}{I_0}\right)$$
$$\text{dB}(6.3\times10^{18} I_0) = 10\log\left(\frac{6.3\times10^{18} I_0}{I_0}\right)$$
$$= 10\log(6.3\times10^{18})$$
$$\approx 187.993$$

The decibel level of the whale is approximately 188 dB.

27. $r_1 = 1.2589^{90} \quad r_2 = 1.2589^{45}$

$$\frac{r_1}{r_2} = \frac{1.2589^{90}}{1.2589^{45}}$$
$$= 1.2589^{90-45} = 1.2589^{45}$$
$$\approx 31{,}594.065$$

No, the intensity of the dog's bark is not twice that of the cat's meow. It is 31,594 times.

28. This is exponential decay and $r = 0.30$. Therefore $p = 1 - 0.30 = 0.70$. Since the rate is every 50 seconds, $T = t/50$. So the exponential function is

$$A = A_0(p)^t$$
$$A = 5000(0.7)^{t/50}$$

Now solve for t when $A = 100$.

$$100 = 5000(0.7)^{t/50}$$
$$\frac{t}{50} = \frac{\log\left(\frac{100}{5000}\right)}{\log(0.7)}$$
$$t = \frac{50\log\left(\frac{100}{5000}\right)}{\log(0.7)} \approx 548.402$$

There will be 100 liters of air in the chapter after approximately 548 seconds.

29. This is exponential growth and $r = 1.07$.

$$A = A_0(r)^t$$
$$7500 = 5000(1.07)^t$$
$$t = \frac{\log\left(\frac{7500}{5000}\right)}{\log(1.07)} \approx 5.993$$

It will take about 6 years to be worth \$7500.

30. Use the half-life equation and its solution.

$$A = A_0\left(\frac{1}{2}\right)^{t/h}$$
$$20 = 25\left(\frac{1}{2}\right)^{4/h}$$
$$h = 4\cdot\frac{\log\left(\frac{1}{2}\right)}{\log\left(\frac{20}{25}\right)} \approx 12.425$$

The half-life is approximately 12.4 hours.

31. $$2(5000) = 5000\left(1+\frac{0.021}{2}\right)^{2t}$$
$$2 = (1.0105)^{2t}$$
$$\log 2 = \log(1.0105)^{2t}$$
$$\log 2 = 2t\log(1.0105)$$
$$t = \frac{\log 2}{2\log(1.0105)} \approx 33$$

It will take approximately 33 years for the value of the account to double in value.

32. a. No, $y = \frac{2}{x^2} - \frac{3}{x} + 4$ is not a quadratic function since the variable is in the denominator.

b. Yes, $s = \frac{1}{2}t^2 + 3$ is a quadratic function.

$a = \frac{1}{2},\ b = 0,\ c = 3$

c. No, $y = 2x - 3$ is not a quadratic function. It is a linear function.

d. Yes, $y = 1 - x - x^2$ is a quadratic function.

Rewrite as $y = -x^2 - x + 1$.

$a = -1,\ b = -1,\ c = 1$

33. a. Since $a = 1 > 0$, the graph opens up.

$$x = -\frac{b}{2a} = -\frac{-8}{2(1)} = 4$$
$$f(4) = (4)^2 - 8(4) - 6 = -22$$

The vertex is (4,–22).

b. Since $a = -1 < 0$, the graph opens down.

$$x = -\frac{b}{2a} = -\frac{3}{2(-1)} = \frac{3}{2} = 1.5$$
$$f\left(\frac{3}{2}\right) = -\left(\frac{3}{2}\right)^2 + 3\left(\frac{3}{2}\right) = \frac{9}{4} = 2.25$$

The vertex is $\left(\frac{3}{2},\ \frac{9}{4}\right)$ or (1.5, 2.25).

c. Since $a = -2 < 0$, the graph opens down.

$$z = -\frac{b}{2a} = -\frac{-4}{2(-2)} = -1$$
$$f(-1) = -2(-1)^2 - 4(-1) - 3 = -1$$

The vertex is (–1,–1).

d. Since $a = 1 > 0$, the graph opens up.

$$x = -\frac{b}{2a} = -\frac{-1}{2(1)} = \frac{1}{2} = 0.5$$
$$f\left(\frac{1}{2}\right) = 3 - \left(\frac{1}{2}\right) + \left(\frac{1}{2}\right)^2 = \frac{11}{4} = 2.75$$

The vertex is $\left(\frac{1}{2},\ \frac{11}{4}\right)$ or (0.5, 2.75).

34. a. To find the x-intercepts, let $y = 0$ and solve for x.

$$0 = x^2 - 2x - 8$$
$$0 = (x-4)(x+2)$$
$$x - 4 = 0 \quad \text{or} \quad x + 2 = 0$$
$$x = 4 \qquad\qquad x = -2$$

To find the y-intercept, let $x = 0$ and solve for y.

$$y = 0^2 - 2(0) - 8 = -8$$

The x-intercepts are (–2, 0) and (4, 0). The y-intercept is (0,–8).

b. To find the x-intercepts, let $y = 0$ and solve for x.

$$0 = -x^2 + 3x + 18$$
$$0 = -(x^2 - 3x - 18)$$
$$0 = -(x-6)(x+3)$$
$$x - 6 = 0 \quad \text{or} \quad x + 3 = 0$$
$$x = 6 \qquad\qquad x = -3$$

To find the y-intercept, let $x = 0$ and solve for y.

$$y = -0^2 + 3(0) + 18 = 18$$

The x-intercepts are (–3, 0) and (6, 0). The y-intercept is (0, 18).

35. Find t when $y = 0$.

$0 = -16t^2 + 64.952t + 5$

$t = \dfrac{-64.952 \pm \sqrt{(64.952)^2 - 4(-16)(5)}}{2(-16)}$

$t = \dfrac{-64.952 \pm \sqrt{4538.762304}}{-32}$

$t \approx \dfrac{-64.952 \pm 67.370}{-32}$

$t \approx \dfrac{-64.952 + 67.37}{-32}$ or $t \approx \dfrac{-64.952 - 67.37}{-32}$

≈ -0.076 ≈ 4.135

The hang time is approximately 4.1 seconds.

36. Find x when $y = 0$.

$0 = -0.0113778x^2 + 1.73205x + 5$

Then $x =$

$\dfrac{-1.73205 \pm \sqrt{(1.73205)^2 - 4(-0.0113778)(5)}}{2(-0.0113778)}$

Simplifying, we have

$x \approx \dfrac{-1.73205 \pm 1.7965}{-0.0227556}$

$x \approx \dfrac{-1.73205 + 1.7965}{-0.0227556}$ or $x \approx \dfrac{-1.73205 - 1.7965}{-0.0227556}$

≈ -2.832 ≈ 155.063

The distance is approximately 155 feet.

37. a. Find the minimum or maximum.

$x = -\dfrac{b}{2a} = -\dfrac{4}{2(1)} = -2$

$y(-2) = (-2)^2 + 4(-2) - 1 = -5$

Since $a = 1 > 0$, the vertex is a minimum.
The minimum value is –5.

b. Find the minimum or maximum.

$x = -\dfrac{b}{2a} = -\dfrac{6}{2(-2)} = \dfrac{3}{2} = 1.5$

$y\left(\dfrac{3}{2}\right) = -2\left(\dfrac{3}{2}\right)^2 + 6\left(\dfrac{3}{2}\right) + 3 = \dfrac{15}{2} = 7.5$

Since $a = -2 < 0$, the vertex is a maximum.

The maximum value is $\dfrac{15}{2}$ or 7.5.

38. $s = -16t^2 + 64t + 6$

a. $t = -\dfrac{b}{2a} = -\dfrac{64}{2(-16)} = 2$ seconds

b. $s(2) = -16(2)^2 + 64(2) + 6 = 70$ feet

39. Find the maximum area.
Using the formula for perimeter for three sides,

$P = 2w + l \Rightarrow 72 = 2x + y$

$y = 72 - 2x$

Using the formula for area, $A = xy$. Then

$A(x) = x(72 - 2x)$

$A(x) = -2x^2 + 72x$

$x = -\dfrac{b}{2a} = -\dfrac{72}{2(-2)} = 18$

$y = 72 - 2(18) = 36$

18 ft by 36 ft

40. Let x = the number of washing machines sold.

$R(x) = xp = x(2700 - 30x) = -30x^2 + 2700x$

$P(x) = R(x) - C(x)$

$= (-30x^2 + 2700x) - (600x + 2200)$

$= -30x^2 + 2100x - 2200$

$x = -\dfrac{b}{2a} = -\dfrac{2100}{2(-30)} = 35$

$P(35) = -30(35)^2 + 2100(35) - 2200$

$= \$34,550$

35 washing machines should be sold to earn a maximum profit of $34,550.

Chapter 7: Introduction to Probability

THINK ABOUT IT, SECTION 7.1

1. An experiment is an activity with an observable outcome.

2. The sample space is the set of all possible outcomes of an experiment.

3. The sample space is the set of all possible outcomes of an experiment. An event contains one or more of the possible outcomes of an experiment, so an event is a subset of the sample space.

4. More. Each stage of an experiment (after the first) performed without replacement will have fewer possible outcomes than the preceding stage. Performed with replacement, each stage of the experiment has the same number of outcomes.

SECTION EXERCISES 7.1

1. {0, 2, 4, 6, 8}

3. {Monday, Tuesday, Wednesday, Thursday, Friday, Saturday, Sunday}

5.

	H	T
H	HH	HT
T	TH	TT

{HH, TT, HT, TH}

7.

	H	T
1	1H	1T
2	2H	2T
3	3H	3T
4	4H	4T
5	5H	5T
6	6H	6T

{1H, 2H, 3H, 4H, 5H, 6H,1T, 2T, 3T, 4T, 5T, 6T}

9. $\{S_1E_1D_1, S_1E_1D_2, S_1E_2D_1, S_1E_2D_2, S_1E_3D_1, S_1E_3D_2, S_2E_1D_1, S_2E_1D_2, S_2E_2D_1, S_2E_2D_2, S_2E_3D_1, S_2E_3D_2\}$

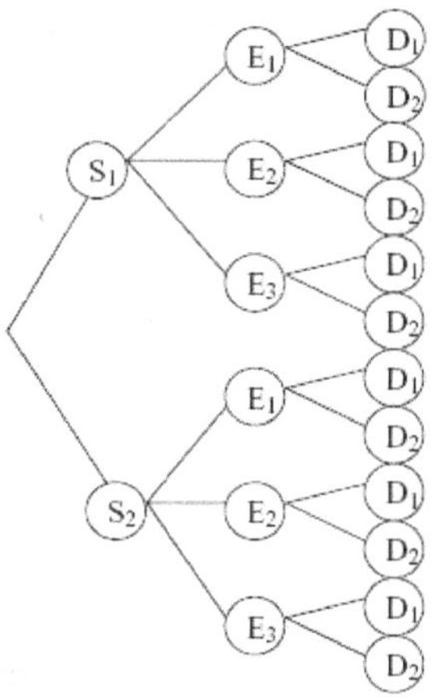

11. {ABCD, ABDC, ACBD, ACDB, ADBC, ADCB}

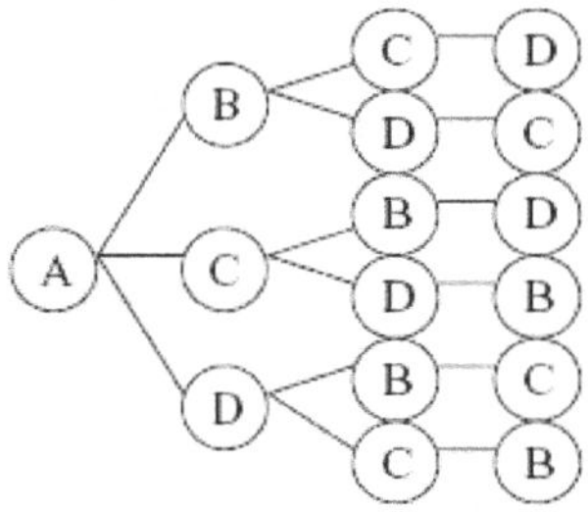

13. Any of the 4 numbers can be used in the first digit, so $n_1 = 4$. Because each number can be used only once, $n_2 = 3$. $4 \cdot 3 = 12$ two digit numbers.

15. There are four choices on each question, so there are $4 \cdot 4 \cdot 4 \cdot 4 \cdot 4 \cdot 4 \cdot 4 \cdot 4 \cdot 4 \cdot 4 \cdot 4 \cdot 4 \cdot 4 \cdot 4 \cdot 4 \cdot 4 \cdot 4 \cdot 4 \cdot 4 \cdot 4 = 4^{20}$ possible ways to complete the test.

17. There are 7 numbers that can be used in the first digit, so $n_1 = 7$. There are 10 numbers that can be used in the last three digits, so $n_2 = 10$, $n_3 = 10$ and $n_4 = 10$. By the counting principle, there are $7 \cdot 10 \cdot 10 \cdot 10 = 7000$ possible four-digit numbers.

19. Because a number starting with 0 is not two-digits, there are 9 numbers that can be used in the first digit, so $n_1 = 9$. There are 10 numbers that can be used in the last digit, so $n_2 = 10$. By the counting principle, there are $9 \cdot 10 = 90$ possible two-digit numbers.

21. Because a number starting with 0 is not two- digits, there are 9 numbers that can be used in the first digit, so $n_1 = 9$. Because a number divisible by 5 must end in 0 or 5, there are 2 numbers that can be used in the last digit, so $n_2 = 2$. By the counting principle, there are $9 \cdot 2 = 18$ possible two-digit numbers divisible by 5.

23. Because a number with 4 or greater in the first place is greater than 37, there are 6 numbers that can be used in the first digit, so $n_1 = 6$. For numbers starting with 4 or greater, any number may be in the second digit, so $n_2 = 10$. Additionally, some numbers may start with 3, so for these cases $n_1 = 1$. When paired with 3, 8 and 9 are the only numbers that make a number greater than 37, so $n_2 = 2$. There are $6 \cdot 10 = 60$ numbers and $1 \cdot 2 = 2$ numbers or $60 + 2 = 62$ possible two-digit numbers greater than 37.

25. Any one of the cards can be placed in box A, so $n_1 = 4$. Because a card cannot be placed in more than one box, there are 3 cards that can be placed in box B, so $n_2 = 3$. Similarly, there are 2 cards that can be placed in box C and 1 card in box D. By the counting principle, there are $4 \cdot 3 \cdot 2 \cdot 1 = 24$ possible placements.

27. Using the sample space found in Exercise 26, there are 15 elements where at least one card is in the box with the corresponding letter.

29. There are thirteen choices in each position, so there are $13 \cdot 13 \cdot 13 \cdot 13 = 13^4$ possible ways to choose the cards.

31. There is 1 choice for each position, so n_1, n_2, n_3, and $n_4 = 1$. By the counting principle, there are $1 \cdot 1 \cdot 1 \cdot 1 = 1$ possible ways to choose the cards.

33. Answers may vary.

Think About It, Section 7.2

1. $0! = 1$

2. A permutation is an arrangement of objects in a definite order.
A combination is a collection of objects in which the order of the objects is not important.

3. a. permutation

   b. permutation

   c. combination

   d. combination

Section Exercises 7.2

1. $8! = 8 \cdot 7 \cdot 6 \cdot 5 \cdot 4 \cdot 3 \cdot 2 \cdot 1 = 40{,}320$

3. Evaluating:
$9! - 5! = (9 \cdot 8 \cdot 7 \cdot 6 \cdot 5 \cdot 4 \cdot 3 \cdot 2 \cdot 1) - (5 \cdot 4 \cdot 3 \cdot 2 - 1)$
$= 362{,}880 - 120 = 362{,}760$

5. $(8-3)! = 5! = 5 \cdot 4 \cdot 3 \cdot 2 \cdot 1 = 120$

7. Evaluating:
$$P(8,\ 5) = \frac{8!}{(8-5)!} = \frac{8!}{3!} = \frac{8 \cdot 7 \cdot 6 \cdot 5 \cdot 4 \cdot 3!}{3!} = 8 \cdot 7 \cdot 6 \cdot 5 \cdot 4 = 6720$$

9. Evaluating:
$$P(9,\ 7) = \frac{9!}{(9-7)!} = \frac{9!}{2!} = \frac{9 \cdot 8 \cdot 7 \cdot 6 \cdot 5 \cdot 4 \cdot 3 \cdot 2!}{2!} = 9 \cdot 8 \cdot 7 \cdot 6 \cdot 5 \cdot 4 \cdot 3 = 181{,}440$$

11. $P(8,\ 0) = \dfrac{8!}{(8-0)!} = \dfrac{8!}{8!} = 1$

13. $$P(8,\ 8) = \frac{8!}{(8-8)!} = \frac{8!}{0!} = \frac{8 \cdot 7 \cdot 6 \cdot 5 \cdot 4 \cdot 3 \cdot 2 \cdot 1}{1} = 8 \cdot 7 \cdot 6 \cdot 5 \cdot 4 \cdot 3 \cdot 2 \cdot 1 = 40{,}320$$

15. Evaluating:
$$P(8,\ 2) \cdot P(5,\ 3) = \frac{8!}{(8-2)!} \cdot \frac{5!}{(5-3)!} = \frac{8 \cdot 7 \cdot 6!}{6!} \cdot \frac{5 \cdot 4 \cdot 3 \cdot 2!}{2!} = 56 \cdot 60 = 3360$$

17. Evaluating:
$$\frac{P(6,\ 0)}{P(6,\ 6)} = \frac{\frac{6!}{(6-0)!}}{\frac{6!}{(6-6)!}} = \frac{\frac{6!}{6!}}{\frac{6!}{0!}} = \frac{1}{\frac{6 \cdot 5 \cdot 4 \cdot 3 \cdot 2 \cdot 1}{1}} = \frac{1}{720}$$

19. Evaluating:
$$C(9,\ 2) = \frac{9!}{2!(9-2)!} = \frac{9!}{2!7!} = \frac{9 \cdot 8 \cdot 7!}{2!7!} = \frac{9 \cdot 8}{2 \cdot 1} = 36$$

21. Evaluating:
$$C(12,\ 0) = \frac{12!}{0!(12-0)!} = \frac{12!}{0!12!} = \frac{1}{1} = 1$$

23. Evaluating:
$$C(6,\ 2) \cdot C(7,\ 3) = \frac{6!}{2!(6-2)!} \cdot \frac{7!}{3!(7-3)!} = \frac{6!}{2!4!} \cdot \frac{7!}{3!4!} = \frac{6 \cdot 5 \cdot 4!}{2!4!} \cdot \frac{7 \cdot 6 \cdot 5 \cdot 4!}{3!4!} = \frac{6 \cdot 5}{2 \cdot 1} \cdot \frac{7 \cdot 6 \cdot 5}{3 \cdot 2 \cdot 1} = 15 \cdot 35 = 525$$

25. Evaluating:

$$\frac{C(10,\ 4)\cdot C(5,\ 2)}{C(15,\ 6)}=\frac{\frac{10!}{4!(10-4)!}\cdot\frac{5!}{2!(5-2)!}}{\frac{15!}{6!(15-6)!}}$$
$$=\frac{\frac{10!}{4!6!}\cdot\frac{5!}{2!3!}}{\frac{15!}{6!9!}}$$
$$=\frac{\frac{10\cdot 9\cdot 8\cdot 7\cdot 6!}{2!18!}\cdot\frac{5\cdot 4\cdot 3!}{2!3!}}{\frac{15\cdot 14\cdot 13\cdot 12\cdot 11\cdot 10\cdot 9!}{6!9!}}$$
$$=\frac{\frac{10\cdot 9\cdot 8\cdot 7}{4\cdot 3\cdot 2\cdot 1}\cdot\frac{5\cdot 4}{2\cdot 1}}{\frac{15\cdot 14\cdot 13\cdot 12\cdot 11\cdot 10}{6\cdot 5\cdot 4\cdot 3\cdot 2\cdot 1}}$$
$$=\frac{210\cdot 10}{5005}=\frac{60}{143}$$

27. Evaluating:

$$\frac{C(9,\ 7)\cdot C(5,\ 3)}{C(14,\ 10)}=\frac{\frac{9!}{7!(9-7)!}\cdot\frac{5!}{3!(5-3)!}}{\frac{14!}{10!(14-10)!}}$$
$$=\frac{\frac{9!}{7!2!}\cdot\frac{5!}{3!2!}}{\frac{14!}{10!4!}}$$
$$=\frac{\frac{9\cdot 8\cdot 7!}{7!2!}\cdot\frac{5\cdot 4\cdot 3!}{3!2!}}{\frac{14\cdot 13\cdot 12\cdot 11\cdot 10!}{10!4!}}$$
$$=\frac{\frac{9\cdot 8}{2\cdot 1}\cdot\frac{5\cdot 4}{2\cdot 1}}{\frac{14\cdot 13\cdot 12\cdot 11}{4\cdot 3\cdot 2\cdot 1}}$$
$$=\frac{36\cdot 10}{1001}=\frac{360}{1001}$$

29. Evaluating:

$$4!\cdot C(10,\ 3)=4!\cdot\frac{10!}{3!(10-3)!}=4!\cdot\frac{10!}{3!7!}$$
$$=4!\cdot\frac{10\cdot 9\cdot 8\cdot 7!}{3!7!}$$
$$=4\cdot 10\cdot 9\cdot 8=2880$$

31. Using the formula:

$$C(7,\ 5)=\frac{7!}{5!(7-5)!}=\frac{7!}{5!2!}$$
$$=\frac{7\cdot 6\cdot 5!}{5!2!}=\frac{7\cdot 6}{2\cdot 1}=21$$

33. Using the formula:

$$C(12,\ 7)=\frac{12!}{7!(12-7)!}=\frac{12!}{7!5!}$$
$$=\frac{12\cdot 11\cdot 10\cdot 9\cdot 8\cdot 7!}{7!5!}$$
$$=\frac{12\cdot 11\cdot 10\cdot 9\cdot 8}{5\cdot 4\cdot 3\cdot 2\cdot 1}=792$$

35. No. If there are only 7 items, there is no way to choose 9 of them.

37. The order in which the songs are played is important, so the number of ways to order the songs is

$$P(5,\ 5)=\frac{5!}{(5-5)!}=\frac{5!}{0!}=\frac{5\cdot 4\cdot 3\cdot 2\cdot 1}{1}$$
$$=5\cdot 4\cdot 3\cdot 2\cdot 1=120$$

39. Because each officer position is different, the order of the selection is important, so the number of ways to make the selection is

$$P(16,\ 4)=\frac{16!}{(16-4)!}=\frac{16!}{12!}$$
$$=\frac{16\cdot 15\cdot 14\cdot 13\cdot 12!}{12!}$$
$$=16\cdot 15\cdot 14\cdot 13=43{,}680$$

41. Because band arrangements have a definite order, they are permutations.
   a. If there are no restrictions on the band order, then the number of different band orders is:

$$P(9,\ 9)=\frac{9!}{(9-9)!}=\frac{9!}{0!}=9!=362{,}880$$

   b. There are 6! ways to arrange the country bands and 3! ways to arrange the rock bands. We must also consider that either the country bands or the rock bands could be arranged at the beginning of the show. There are two ways to do this. By the counting principle, there are $2\cdot 6!\cdot 3!=8640$ arrangements for the bands to play.

43. We are looking for the number of permutations of the characters SOCCER77.
With $n = 8$ (number of characters),
$k_1 = 1$ (number of S's), $k_2 = 1$ (number of O's),
$k_3 = 2$ (number of C's), $k_4 = 1$ (number of E's),
$k_5 = 1$ (number of R's), and
$k_6 = 2$ (number of 7's), we have

$$\frac{8!}{1!1!2!1!1!2!}=10{,}080$$

There are 10,080 possible passwords.

45. The order in which the firefighters are selected is not important, so the number of different teams is

$$C(24,\ 8)=\frac{24!}{8!(24-8)!}=\frac{24!}{8!16!}$$
$$=\frac{24\cdot 23\cdot 22\cdot 21\cdot 20\cdot 19\cdot 18\cdot 17\cdot 16!}{8!16!}$$
$$=\frac{24\cdot 23\cdot 22\cdot 21\cdot 20\cdot 19\cdot 18\cdot 17}{8\cdot 7\cdot 6\cdot 5\cdot 4\cdot 3\cdot 2\cdot 1}$$
$$=735{,}471$$

47. The order of the questions on the test is not important, so the number of different exams is

$$C(7,\ 3)=\frac{7!}{3!(7-3)!}=\frac{7!}{3!4!}$$
$$=\frac{7\cdot6\cdot5\cdot4!}{3!4!}=\frac{7\cdot6\cdot5}{3\cdot2\cdot1}=35$$

49. The order of the choices is not important, so the number of possible ways to choose the women, is $C(8, 3)$ and the number of ways to choose the men is $C(8, 3)$. Therefore, by the counting principle, there are $C(8,3)\cdot C(8,3)$ different committees.

$$C(8,\ 3)\cdot C(8,\ 3)=\frac{8!}{3!(8-3)!}\cdot\frac{8!}{3!(8-3)!}$$
$$=\frac{8!}{3!5!}\cdot\frac{8!}{3!5!}$$
$$=\frac{8\cdot7\cdot6\cdot5!}{5!3!}\cdot\frac{8\cdot7\cdot6\cdot5!}{3!5!}$$
$$=\frac{8\cdot7\cdot6}{3\cdot2\cdot1}\cdot\frac{8\cdot7\cdot6}{3\cdot2\cdot1}$$
$$=56\cdot56=3136$$

51. Because the choice of team A playing team B is the same as the choice of team B playing team A, the order of selection is not important, so the number of different ways to match the teams is

$$C(4,\ 2)=\frac{4!}{2!(4-2)!}=\frac{4!}{2!2!}$$
$$=\frac{4\cdot3\cdot2!}{2!2!}=\frac{4\cdot3}{2\cdot1}=6$$

Since the teams play each other twice, the number of games is $6\cdot2=12$.

53. There are 6 points and we must choose 2 to form a diagonal. Because order is not important, we find $C(6, 2)$.

$$C(6,\ 2)=\frac{6!}{2!(6-2)!}=\frac{6!}{2!4!}$$
$$=\frac{6\cdot5\cdot4!}{2!4!}=\frac{6\cdot5}{2\cdot1}=15$$

Because 6 of the lines made are the sides of the hexagon, the number of diagonals is $15-6=9$.

55. The order in which players are selected is not important, so the number of different teams is

$$C(18,\ 9)=\frac{18!}{9!(18-9)!}=\frac{18!}{9!9!}$$
$$=\frac{18\cdot17\cdot16\cdot15\cdot14\cdot13\cdot12\cdot11\cdot10!}{9!9!}$$
$$=\frac{18\cdot17\cdot16\cdot15\cdot14\cdot13\cdot12\cdot11\cdot10}{9\cdot8\cdot7\cdot6\cdot5\cdot4\cdot3\cdot2\cdot1}$$
$$=48{,}620$$

57. The order in which the flags are raised is important, so the number of different signals possible is

$$P(4,\ 4)=\frac{4!}{(4-4)!}=\frac{4!}{0!}=\frac{4\cdot3\cdot2\cdot1}{1}$$
$$=4\cdot3\cdot2\cdot1=24$$

59. We are looking for the number of permutations of the letters *committee*. With
$n=9$ (number of letters),
$k_1=1$ (number of *c*'s), $k_2=1$ (number of *o*'s),
$k_3=2$ (number of *m*'s), $k_4=1$ (number of *i*'s),
$k_5=2$ (number of *t*'s), $k_6=2$ (number of *e*'s),
we have

$$\frac{9!}{1!1!2!1!2!2!}=45{,}360$$

There are 45,360 different arrangements possible.

61. The order in which the heads and tails are chosen is not important, so the number of distinct arrangements is

$$C(10,\ 5)=\frac{10!}{5!(10-5)!}=\frac{10!}{5!5!}$$
$$=\frac{10\cdot9\cdot8\cdot7\cdot6\cdot5!}{5!5!}$$
$$=\frac{10\cdot9\cdot8\cdot7\cdot6}{5\cdot4\cdot3\cdot2\cdot1}=252$$

63. Because the group arrangements have a definite order, they are permutations.
There are 3! ways to arrange the first group, 4! ways to arrange the second group, and 2! ways to arrange the third group. We must also consider that either the first, second, or third group could be arranged at the beginning of the show. There are 3! ways to do this.
By the counting principle, there are $3!\cdot3!\cdot4!\cdot2!=1728$ possible orders of the groups.

65. There are $C(4, 4)$ ways of choosing four aces from four aces and 48 choices for the fifth card in the hand. By the counting principle, the number of hands containing 4 aces is

$$C(4,\ 4)\cdot48=\frac{4!}{4!(4-4)!}\cdot48=\frac{4!}{4!0!}\cdot48$$
$$=\frac{1}{1}\cdot48=48$$

67. There are $C(4, 3)$ ways of choosing three jacks from four jacks and $C(48, 2)$ ways of choosing the other 2 cards from the remaining 48 cards. By the counting principle, the number of hands containing exactly three jacks is

$$C(4,\ 3)\cdot C(48,\ 2)=\frac{4!}{3!(4-3)!}\cdot\frac{48!}{2!(48-2)!}$$
$$=\frac{4!}{3!1!}\cdot\frac{48!}{2!46!}$$
$$=\frac{4\cdot3!}{3!1!}\cdot\frac{48\cdot47\cdot46!}{2!46!}$$
$$=\frac{4}{1}\cdot\frac{48\cdot47}{2\cdot1}$$
$$=4\cdot1128=4512$$

69. There are $C(4, 2)$ ways of choosing two sevens from four sevens and $C(48, 3)$ ways of choosing the other 3 cards from the remaining 48 cards. By the counting principle, the number of hands containing exactly two sevens is

$$\begin{aligned} C(4, 2)\cdot C(48, 3) &= \frac{4!}{2!(4-2)!}\cdot\frac{48!}{3!(48-3)!} \\ &= \frac{4!}{2!2!}\cdot\frac{48!}{2!45!} \\ &= \frac{4\cdot 3\cdot 2!}{2!2!}\cdot\frac{48\cdot 47\cdot 46\cdot 45!}{2!45!} \\ &= \frac{4\cdot 3}{2\cdot 1}\cdot\frac{48\cdot 47\cdot 46}{3\cdot 2\cdot 1} \\ &= 6\cdot 17,296 = 103,776 \end{aligned}$$

THINK ABOUT IT, SECTION 7.3

1. No, all probabilities are between 0 and 1, inclusive.

2. Yes, each pencil has the same chance of being selected.

3. No, the probability of, for instance, a sum of 7 is not the same as the probability of a sum of 11.

4. $\frac{1}{2}$

5. Less than. There are fewer red balls.

SECTION EXERCISES 7.3

1. $S = \{$HHH, HHT, HTH, HTT, THH, THT, TTH, TTT$\}$

3. $S = \{$Nov. 1, Nov. 2, Nov. 3, Nov. 4, Nov. 5, Nov. 6, Nov. 7, Nov. 8, Nov. 9, Nov. 10, Nov. 11, Nov. 12, Nov. 13, Nov. 14$\}$

5. $S = \{$Alaska, Alabama, Arizona, Arkansas$\}$

7. $S = \{$BBB, BBG, BGB, BGG, GBB, GBG, GGB, GGG$\}$

9. $S = \{$BGG, GBG, GGB, GGG$\}$

11. $S = \{$BBG, BGB, BGG, GBB, GBG, GGB, GGG$\}$

13. The sample space was found in Exercise 7. The elements in the event that the Lins will have at least two girls was found in Exercise 9. Then the probability is $P(E) = \frac{n(E)}{n(S)} = \frac{4}{8} = \frac{1}{2}$.

15. The sample space was found in Exercise 7. The elements in the event that the Lins will have at least one girl are $\{$BBG, BGB, GBB, BGG, GBG, GGB, GGG$\}$. Then the probability is $P(E) = \frac{n(E)}{n(S)} = \frac{7}{8}$.

17. The sample space for tossing a coin four times is $\{$HHHH, HHHT, HHTH, HHTT, HTHH, HTHT, HTTH, HTTT, THHH, THHT, THTH, THTT, TTHH, TTHT, TTTH, TTTT$\}$. The elements in the event that one head and three tails are tossed are $\{$HTTT, THTT, TTHT, TTTH$\}$. Then the probability is $P(E) = \frac{n(E)}{n(S)} = \frac{4}{16} = \frac{1}{4}$.

19. The sample space for tossing a dodecahedral die is $\{1, 2, 3, 4, 5, 6, 7, 8, 9, 10, 11, 12\}$. The element in the event that the number on the upward face is 12 is $\{12\}$. Then the probability is $P(E) = \frac{n(E)}{n(S)} = \frac{1}{12}$.

21. The sample space for tossing a dodecahedral die was found in Exercise 19. The elements in the event that the number on the upward face is divisible by 4 are $\{4, 8, 12\}$. Then the probability is $P(E) = \frac{n(E)}{n(S)} = \frac{3}{12} = \frac{1}{4}$.

23. The sample space for tossing two six-sided dice was found in the text. The dice must be considered as distinct, so there are 36 possible outcomes. The elements in the event that the sum of the pips on the upward faces is 6 are

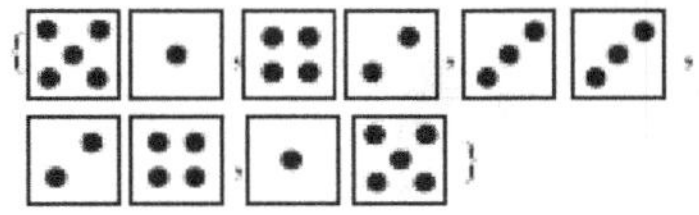

Then the probability is $P(E) = \frac{n(E)}{n(S)} = \frac{5}{36}$.

25. The sample space for tossing two six-sided dice was found in the text. The dice must be considered as distinct, so there are 36 possible outcomes. The element in the event that the sum of the pips on the upward faces is 2 is

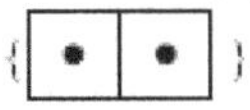

Then the probability is $P(E) = \frac{n(E)}{n(S)} = \frac{1}{36}$.

27. The sample space for tossing two six-sided dice was found in the text. The dice must be considered as distinct, so there are 36 possible outcomes. There are no elements in the event that the sum of the pips on the upward faces is 1.

    Then the probability is $P(E)=\frac{n(E)}{n(S)}=\frac{0}{36}=0.$

29. The sample space for tossing two six-sided dice was found in the text. The dice must be considered as distinct, so there are 36 possible outcomes. The elements in the event that the sum of the pips on the upward faces is at least 10 are (6 and 6, 6 and 5, 6 and 4, 5 and 6, 5 and 5, 4 and 6}. Then the probability is

    $P(E)=\frac{n(E)}{n(S)}=\frac{6}{36}=\frac{1}{6}.$

31. The sample space for tossing two six-sided dice was found in the text. The dice must be considered as distinct, so there are 36 possible outcomes. The elements in the event that the sum of the pips on the upward faces is an even number are {6 and 6, 6 and 4, 6 and 2, 5 and 5, 5 and 3, 5 and 1, 4 and 6, 4 and 4, 4 and 2, 3 and 5, 3 and 3, 3 and 1, 2 and 6, 2 and 4, 2 and 2, 1 and 5, 1 and 3, 1 and 1}.
    Then the probability is

    $P(E)=\frac{n(E)}{n(S)}=\frac{18}{36}=\frac{1}{2}.$

33. The sample space for tossing two six-sided dice was found in the text. The dice must be considered as distinct, so there are 36 possible outcomes. The elements in the event that both dice show the same number of pips are {6 and 6, 5 and 5, 4 and 4, 3 and 3, 2 and 2, 1 and 1}.
    Then the probability is

    $P(E)=\frac{n(E)}{n(S)}=\frac{6}{36}=\frac{1}{6}.$

35. Because there are 52 cards in a standard deck of playing cards, there are 52 items in the sample space. The elements in the event that a nine is drawn are {9 hearts, 9 diamonds, 9 spades, 9 clubs}. Then the probability is

    $P(E)=\frac{n(E)}{n(S)}=\frac{4}{52}=\frac{1}{13}.$

37. Because there are 52 cards in a standard deck of playing cards, there are 52 items in the sample space. The elements in the event that a card between 5 and 9 inclusive is drawn, which includes the 5, 6, 7, 8, and 9 in each of four suits. Then the probability is

    $P(E)=\frac{n(E)}{n(S)}=\frac{15}{52}=\frac{5}{13}.$

39. Let E be the event that the voter is a Democrat.

    Then $P(E)=\frac{1267}{3228}\approx 0.39.$

41. Let E be the event that the voter is 50 years old or older. Then

    $P(E)=\frac{804}{3228}=\frac{67}{269}\approx 0.25.$

43. Let E be the event that the voter is an Independent. Then

    $P(E)=\frac{150}{3228}=\frac{25}{538}\approx 0.05.$

45. Let E be the event that the respondent did not complete high school. Then

    $P(E)=\frac{52}{850}=\frac{26}{425}.$

47. Let E be the event that the respondent has a Ph.D. or professional degree. Then

    $P(E)=\frac{36}{850}=\frac{18}{425}.$

49. Let E be the event that the respondent earns from \$36,000 to \$45,999 annually. Then

    $P(E)=\frac{58}{293}.$

51. Let E be the event that the respondent earns at least \$80,000 annually. There are 22 + 14 = 36 people in the survey that earn at least \$80,000 annually. Then

    $P(E)=\frac{36}{293}.$

53. To have white flowers, the plant must be *rr*. From the table, only one of the four possible genotypes is *rr*, so the probability that an offspring will have white flowers is $\frac{1}{4}$.

55. Make a Punnett square.

Parents	*E*	*E*
e	*Ee*	*Ee*
e	*Ee*	*Ee*

    To have red eyes, the mouse must be *ee*. From the table, none of the four possible genotypes is *ee*, so the probability that an offspring will have red eyes is 0.

57. Use the formula for calculating the probability of an event when the odds in favor are known.

    $P(E)=\frac{a}{a+b}=\frac{1}{1+2}=\frac{1}{3}$

59. Use the formula for calculating the probability of an event when the odds in favor are known.

    $P(E)=\frac{a}{a+b}=\frac{3}{3+7}=\frac{3}{10}$

61. Use the formula for calculating the probability of an event when the odds in favor are known.

$$P(E) = \frac{a}{a+b} = \frac{8}{8+5} = \frac{8}{13}$$

63. The odds in favor are

$$\frac{P(E)}{1-P(E)} = \frac{0.2}{1-0.2} = \frac{0.2}{0.8} = \frac{1}{4}$$

The odds in favor are 1 to 4.

65. The odds in favor are

$$\frac{P(E)}{1-P(E)} = \frac{0.375}{1-0.375} = \frac{0.375}{0.625} = \frac{3}{5}$$

The odds in favor are 3 to 5.

67. The odds in favor are

$$\frac{P(E)}{1-P(E)} = \frac{0.55}{1-0.55} = \frac{0.55}{0.45} = \frac{11}{9}$$

The odds in favor are 11 to 9.

69. There are 30 answers total. Two are desirable and the other 28 are undesirable. The odds of choosing a Daily Double are 2:28 or 1:14, or 1 to 14.

71. Let E be the event of rolling an even number. Because there are three ways to roll an even number, there are three favorable outcomes, leaving three unfavorable outcomes. Odds in favor of $E = \frac{3}{3} = \frac{1}{1}$.

The odds in favor are 1 to 1.

73. Let E be the event of tossing heads four times. Refer to Exercise 16. Because there is one way to toss head four times, there is one favorable outcome, leaving 15 unfavorable outcomes.

Odds against $E = \frac{15}{1}$

The odds against are 15 to 1.

75. Let E be the event of an earthquake of magnitude 7.5 or greater occurring in California within any one year

Because the odds against are 48 to 49, the odds in favor are 49 to 48.

Use the formula for calculating the probability of an event when the odds in favor are known.

$$P(E) = \frac{a}{a+b} = \frac{49}{49+48} = \frac{49}{97}$$

77. Because the odds against are 8 to 3, the odds in favor are 3 to 8. Use the formula for calculating the probability of an event when the odds in favor are known.

$$P(E) = \frac{a}{a+b} = \frac{3}{3+8} = \frac{3}{11}$$

THINK ABOUT IT, SECTION 7.4

1. Yes, a number cannot be both even and odd.

2. No, 3, 5, and 7 are both odd numbers and prime numbers.

3. The complement of the event is $1 - 0.3 = 0.7$.

4. L = {HHHT, HHTH, HTHH, THHH, HHHH}

5. M = {HHTT, HTHT, HTTH, TTHH, THHT, THTH, HTTT, THTT, TTHT, TTTH, TTTT}

SECTION EXERCISES 7.4

1. Let A = {♥4, ♥4, ♦4, ♣4} and B = {♠A, ♥A, ♦A, ♣A}. The events in A and B cannot occur at the same time, so they are mutually exclusive. There are 52 cards in a standard deck, thus $n(S) = 52$. The probability is

$$P(A \text{ or } B) = P(A) + P(B) = \frac{1}{13} + \frac{1}{13} = \frac{2}{13}.$$

3. Let A be the event of rolling a 2, and let B be the event of rolling a 10. The events in A and B cannot occur at the same time, so they are mutually exclusive. From the sample space given in the text,

$$P(A) = \frac{1}{36} \text{ and } P(B) = \frac{3}{36} = \frac{1}{12}.$$

Because A and B are mutually exclusive events,

$$P(A \text{ or } B) = P(A) + P(B) = \frac{1}{36} + \frac{1}{12} = \frac{4}{36} = \frac{1}{9}.$$

5.
$$\begin{aligned} P(A \text{ or } B) &= P(A) + P(B) - P(A \text{ and } B) \\ &= 0.2 + 0.5 - 0.1 \\ &= 0.6 \end{aligned}$$

7.
$$\begin{aligned} P(A \cap B) &= P(A) + P(B) - P(A \cup B) \\ &= 0.3 + 0.8 - 0.9 \\ &= 0.2 \end{aligned}$$

9. Let A be the event of the number being more than 6, and let B be the event of the number being odd. The sample space consists of 10 numbers. $n(A) = 4$, $n(B) = 5$ and $n(A) = 4$, $n(B) = 5$ and $n(A \text{ and } B) = 2$.

$$\begin{aligned} P(A \text{ or } B) &= P(A) + P(B) - P(A \text{ and } B) \\ &= \frac{4}{10} + \frac{5}{10} - \frac{2}{10} = \frac{7}{10}. \end{aligned}$$

11. Let A be the event of the number being even, and let B be the event of the number being prime. The sample space consists of 10 numbers. $n(A) = 5$, $n(B) = 4$ and $n(A \text{ and } B) = 1$.

$$\begin{aligned} P(A \text{ or } B) &= P(A) + P(B) - P(A \text{ and } B) \\ &= \frac{4}{10} + \frac{5}{10} - \frac{1}{10} = \frac{4}{5}. \end{aligned}$$

13. Let A be the event of rolling a 6, and let B be the event of rolling doubles. From the figure in in the text, there are 36 elements in the sample space. $n(A) = 5$, $n(B) = 6$ and $n(A \text{ and } B) = 1$.

$$P(A \text{ or } B) = P(A) + P(B) - P(A \text{ and } B)$$
$$= \frac{5}{36} + \frac{6}{36} - \frac{1}{36} = \frac{10}{36} = \frac{5}{18}.$$

15. Let A be the event of rolling an even number, and let B be the event of rolling doubles. From the figure in the text, there are 36 elements in the sample space. $n(A) = 18$, $n(B) = 6$ and $n(A \text{ and } B) = 6$.

$$P(A \text{ or } B) = P(A) + P(B) - P(A \text{ and } B)$$
$$= \frac{18}{36} + \frac{6}{36} - \frac{6}{36} = \frac{18}{36} = \frac{1}{2}$$

17. Let A be the event of rolling an odd number, and let B be the event of rolling a number less than 6. From the figure in the text, there are 36 elements in the sample space. $n(A) = 18$, $n(B) = 10$ and $n(A \text{ and } B) = 6$.

$$P(A \text{ or } B) = P(A) + P(B) - P(A \text{ and } B)$$
$$= \frac{18}{36} + \frac{10}{36} - \frac{6}{36} = \frac{22}{36} = \frac{11}{18}$$

19. Let A be the event of drawing an 8, and let B be the event of drawing a spade. The sample space consists of 52 cards. $n(A) = 4$, $n(B) = 13$ and $n(A \text{ and } B) = 1$.

$$P(A \text{ or } B) = P(A) + P(B) - P(A \text{ and } B)$$
$$= \frac{4}{52} + \frac{13}{52} - \frac{1}{52} = \frac{16}{52} = \frac{4}{13}$$

21. Let A be the event of drawing a jack, and let B be the event of drawing a face card. The sample space consists of 52 cards. $n(A) = 4$, $n(B) = 12$ and $n(A \text{ and } B) = 4$.

$$P(A \text{ or } B) = P(A) + P(B) - P(A \text{ and } B)$$
$$= \frac{4}{52} + \frac{12}{52} - \frac{4}{52} = \frac{12}{52} = \frac{3}{13}$$

23. Let A be the event of drawing a diamond, and let B be the event of drawing a black card. The sample space consists of 52 cards. $n(A) = 13$, $n(B) = 26$ and $n(A \text{ and } B) = 0$.

$$P(A \text{ or } B) = P(A) + P(B)$$
$$= \frac{13}{52} + \frac{26}{52} = \frac{39}{52} = \frac{3}{4}$$

25. Let $A = \{\text{people aged 26-34}\}$, and let $B = \{\text{people employed part-time}\}$. Then, from the table, $n(A) = 348 + 67 + 27 = 442$, $n(B) = 164 + 203 + 67 + 179 + 162 = 775$ and $n(A \text{ and } B) = 67$. The total number of people represented in the table is 3179.

$$P(A \text{ or } B) = P(A) + P(B) - P(A \text{ and } B)$$
$$= \frac{442}{3179} + \frac{775}{3179} - \frac{67}{3179} = \frac{1150}{3179}$$

27. Let $A = \{\text{people under 18}\}$, and let $B = \{\text{people employed part-time}\}$. Then, from the table, $n(A) = 24 + 164 + 371 = 559$, $n(B) = 164 + 203 + 67 + 179 + 162 = 775$ and $n(A \text{ and } B) = 164$. The total number of people represented in the table is 3179.

$$P(A \text{ or } B) = P(A) + P(B) - P(A \text{ and } B)$$
$$= \frac{559}{3179} + \frac{775}{3179} - \frac{164}{3179} = \frac{1170}{3179}$$

29. If E is the event of winning the contest, then E^C is the event of not winning the contest, and $P(E^C) = 1 - P(E) = 1 - 0.04 = 0.96$

31. If E is the event of a senior being drafted by a NBA team, then E^C is the event of not being drafted on a NBA team, and

$$P(E^C) = 1 - P(E) = 1 - \frac{1}{75} = \frac{74}{75} \approx 98.7\%$$

33. If E is the event of tossing a 7, then E^C is the event of not tossing a 7, and

$$P(E^C) = 1 - P(E) = 1 - \frac{6}{36} = \frac{30}{36} = \frac{5}{6}$$

35. If E is the event of tossing a sum less than 4, then E^C is the event of tossing a sum of at least four, and

$$P(E^C) = 1 - P(E) = 1 - \frac{3}{36} = \frac{33}{36} = \frac{11}{12}$$

37. If E is the event of drawing an ace, then E^C is the event of not drawing an ace, and

$$P(E^C) = 1 - P(E) = 1 - \frac{4}{52} = \frac{48}{52} = \frac{12}{13}$$

39. If E is the event of getting no tails, then E^C is the event of getting at least one tail, and

$$P(E^C) = 1 - P(E) = 1 - \frac{1}{16} = \frac{15}{16}$$

41. Let $E = \{\text{at least one 1}\}$; then $E^C = \{\text{no 1's}\}$. Because on each toss of the die there are six possible outcomes, $n(S) = 6 \cdot 6 \cdot 6 = 216$. On each toss of the die there are five numbers that are not 1's. Therefore, $n(E^C) = 5 \cdot 5 \cdot 5 = 125$.

$$P(E^C) = 1 - P(E) = 1 - \frac{125}{216} = \frac{91}{216} \approx 0.421$$

The probability is 42.1%.

43. Let $E = \{\text{at least one roll of sum of 8}\}$; then $E^C = \{\text{no sum of 8 is rolled}\}$. Using the table in the text, there are 36 possibilities for each toss of the dice. Thus, $n(S) = 36 \cdot 36 \cdot 36 = 46{,}656$. On each toss of the dice there are 31 numbers that do not total 8, so

$$n(E^C) = 31 \cdot 31 \cdot 31 = 29{,}791$$
$$P(E) = 1 - P(E^C) = 1 - \frac{29{,}791}{46{,}656} = \frac{16{,}865}{46{,}656}$$
$$\approx 0.361$$

The probability is 36.1%.

45. Let E = {at least one face card is drawn}; then E^C = {no face card is drawn}. There are 52 possibilities for each card drawn. Thus, $n(S) = 52 \cdot 52 \cdot 52 = 140{,}608$. Each time a card is drawn, there are 40 cards that are not face cards, so $n(E^C) = 40 \cdot 40 \cdot 40 = 64{,}000$.

$$P(E) = 1 - P(E^C) = 1 - \frac{64,000}{140,608}$$
$$= \frac{76,608}{140,608} \approx 0.545$$

The probability is 54.5%.

47. Let E = {at least one e-reader is defective}; then E^C = {no e-readers are defective}. The number of elements in the sample space is

$$n(S) = C(30,\ 12) = \frac{30!}{12!(30-12)!}$$
$$= \frac{30!}{12!18!} = 86,493,225.$$

To find the number of outcomes that contain no defective e-readers , all of them must come from the 26 nondefective e-readers . The number of elements is

$$n(E^C) = C(26,\ 12) = \frac{26!}{12!(26-12)!}$$
$$= \frac{26!}{12!14!} = 9,657,700$$
$$P(E) = 1 - P(E^C) = 1 - \frac{9,657,700}{86,493,225}$$
$$= \frac{76,835,525}{86,493,22} \approx 0.888$$

The probability is 88.8%.

49. Let E = {at least one employee wins a prize}; then E^C = {no employee wins a prize}. The number of elements in the sample space is

$$n(S) = C(42,\ 3) = \frac{42!}{3!(42-3)!}$$
$$= \frac{42!}{3!39!} = 11,480.$$

To find the number of outcomes where no employee wins, all 3 must come from the 39 non-employee cards. The number of elements is

$$n(E^C) = C(39,\ 3) = \frac{39!}{3!(39-3)!}$$
$$= \frac{39!}{3!36!} = 9139.$$
$$P(E) = 1 - P(E^C) = 1 - \frac{9139}{11,480}$$
$$= \frac{2341}{11,480} \approx 0.204$$

The probability is 20.4%.

THINK ABOUT IT, SECTION 7.5

1. The probability that a number is odd given that it is a prime number.

2. Yes, A and B are independent events.

3. Two mutually exclusive events are dependent.

4. The product of the probability of A and the probability of B.

SECTION EXERCISES 7.5

1. $$P(A \mid B) = \frac{P(A \cap B)}{P(B)} = \frac{0.25}{0.4} = 0.625$$
$$P(B \mid A) = \frac{P(A \cap B)}{P(A)} = \frac{0.25}{0.7} \approx 0.357$$

3. $$P(A \mid B) = \frac{P(A \cap B)}{P(B)} = \frac{0.07}{0.18} \approx 0.389$$
$$P(B \mid A) = \frac{P(A \cap B)}{P(A)} = \frac{0.07}{0.61} \approx 0.115$$

5. Let B = {people employed part-time} and A = {people between 35 and 49}. From the table, $n(A \cap B) = 179$, $n(A) = 581 + 179 + 104 = 864$, and $n(S) = 3179$. We have

$$P(B \mid A) = \frac{P(A \cap B)}{P(A)} = \frac{\frac{179}{3179}}{\frac{864}{3179}} = \frac{179}{864}.$$

7. Let B = {people employed full-time} and A = {people between 18 and 49}. From the table, $n(A \cap B) = 185 + 348 + 581 = 1114$, $n(A) = 185 + 203 + 148 + 348 + 67 + 27 + 581 + 179 + 104 = 1842$, and $n(S) = 3179$. We have

$$P(B \mid A) = \frac{P(A \cap B)}{P(A)} = \frac{\frac{1114}{3179}}{\frac{1842}{3179}} = \frac{1114}{1842} = \frac{557}{921}.$$

9. There are 2000 people in the sample space. The number of elements in the event that a participant prefers the Nintendo Wii system is 607. Then the probability is

$$P(E) = \frac{n(E)}{n(S)} = \frac{607}{2000} \approx 0.30.$$

11. Let B = {participants who prefer Sega Dreamcast} and A = {participants between 13 and 24}. From the table, $n(A \cap B) = 92 + 83 = 175$, $n(A) = 105 + 139 + 92 + 113 + 248 + 217 + 83 + 169 = 1166$, and $n(S) = 2000$. We have

$$P(B \mid A) = \frac{P(A \cap B)}{P(A)} = \frac{\frac{175}{2000}}{\frac{1166}{2000}} \approx 0.15.$$

13. Let $B = \{\text{the sum is 8}\}$ and $A = \{\text{the sum is even}\}$. From the table in the text, there are five possible rolls of the dice for which the sum is 8 and the sum is even. So $P(A \cap B) = \frac{5}{36}$. There are 18 possibilities for which the sum is even, so $P(A) = \frac{18}{36} = \frac{1}{2}$. We have

$$P(B \mid A) = \frac{P(A \cap B)}{P(A)} = \frac{\frac{5}{36}}{\frac{1}{2}} = \frac{5}{18}.$$

15. Let $B = \{\text{doubles}\}$ and $A = \{\text{sum less than 7}\}$. From the table in the text, there are three possible rolls of the dice for which the roll is doubles and the sum is less than 7. So $P(A \cap B) = \frac{3}{36} = \frac{1}{12}$. There are 15 possibilities for which the sum is less than 7, so $P(A) = \frac{15}{36} = \frac{5}{12}$. We have

$$P(B \mid A) = \frac{P(A \cap B)}{P(A)} = \frac{\frac{1}{12}}{\frac{5}{12}} = \frac{1}{5}.$$

17. Let $A = \{\text{a face card is drawn first}\}$ and $B = \{\text{a face card is drawn second}\}$. On the first draw, there are 12 face cards in the deck of 52 cards. Therefore, $P(A) = \frac{12}{52} = \frac{3}{13}$. On the second draw, there are only 51 cards remaining and only 11 face cards. Therefore, $P(B \mid A) = \frac{11}{51}$.
We have

$$P(A \cap B) = P(A) \cdot P(B \mid A) = \frac{3}{13} \cdot \frac{11}{51} = \frac{33}{663} \approx 0.050.$$

19. Let $A = \{\text{a spade is drawn first}\}$ and $B = \{\text{a red card is drawn second}\}$. On the first draw, there are 13 spades in the deck of 52 cards. Therefore, $P(A) = \frac{13}{52} = \frac{1}{4}$. On the second draw, there are only 51 cards remaining and 26 red cards. Therefore, $P(B \mid A) = \frac{26}{51}$. We have

$$P(A \cap B) = P(A) \cdot P(B \mid A) = \frac{1}{4} \cdot \frac{26}{51} = \frac{26}{204} \approx 0.127.$$

21. Let $A = \{\text{a blue is selected first}\}$, $B = \{\text{a red is selected second}\}$ and $C = \{\text{a green is selected third}\}$. Then

$$P(A \text{ followed by } B \text{ followed by } C) = P(A) \cdot P(B \mid A) P(C \mid A \text{ and } B) = \frac{12}{57} \cdot \frac{12}{56} \cdot \frac{7}{55} = \frac{6}{1045}.$$

23. Let $A = \{\text{a green is selected first}\}$, $B = \{\text{a green is selected second}\}$ and $C = \{\text{a red is selected third}\}$. Then

$$P(A \text{ followed by } B \text{ followed by } C) = P(A) \cdot P(B \mid A) P(C \mid A \text{ and } B) = \frac{7}{57} \cdot \frac{6}{56} \cdot \frac{12}{55} = \frac{3}{1045}.$$

25. Let $A = \{\text{a red card is dealt first}\}$, $B = \{\text{a black card is dealt second}\}$ and $C = \{\text{a red card is dealt third}\}$. Then

$$P(A \text{ followed by } B \text{ followed by } C) = P(A) \cdot P(B \mid A) P(C \mid A \text{ and } B) = \frac{26}{52} \cdot \frac{26}{51} \cdot \frac{25}{50} = \frac{13}{102}.$$

27. Let $A = \{\text{an ace is dealt first}\}$, $B = \{\text{a face card is dealt second}\}$ and $C = \{\text{an 8 is dealt third}\}$. Then

$$P(A \text{ followed by } B \text{ followed by } C) = P(A) \cdot P(B \mid A) P(C \mid A \text{ and } B) = \frac{4}{52} \cdot \frac{12}{51} \cdot \frac{4}{50} = \frac{8}{5525}.$$

29. Let $A = \{\text{absent the first day}\}$, $B = \{\text{absent the second day}\}$ and $C = \{\text{absent the third day}\}$. Then

$$P(A \text{ followed by } B \text{ followed by } C) = P(A) \cdot P(B \mid A) \cdot P(C \mid A \text{ and } B) = 0.04 \cdot 0.11 \cdot 0.11 = 0.000484.$$

31. Let $A = \{\text{absent the first day}\}$, $B = \{\text{not absent the second day}\}$ and $C = \{\text{absent the third day}\}$. Then

$$P(A \text{ followed by } B \text{ followed by } C) = P(A) \cdot P(B \mid A) \cdot P(C \mid A \text{ and } B) = 0.04 \cdot 0.89 \cdot 0.04 = 0.001424.$$

33. The outcome of the first roll has no effect on the outcome of the second roll so the events are independent.

35. The probability of numbers on the second slip of paper depends on the result of the first slip of paper so the events are not independent.

37. The rolls of a pair of dice are independent. Let $A = \{\text{sum of 8 on the first roll}\}$ and $B = \{\text{sum of 8 on the second roll}\}$.

$$P(A \cap B) = P(A) \cdot P(B) = \frac{5}{36} \cdot \frac{5}{36} = \frac{25}{1296}$$

39. The rolls of a pair of dice are independent. Let $A = \{\text{sum of at least 10 on the first roll}\}$ and $B = \{\text{sum of at least 11 on the second roll}\}$.

$$P(A \cap B) = P(A) \cdot P(B) = \frac{6}{36} \cdot \frac{3}{36} = \frac{18}{1296} = \frac{1}{72}$$

41. The rolls of a pair of dice are independent. Let $A = \{\text{sum of even number on the first roll}\}$ and $B = \{\text{sum of even number on the second roll}\}$.
$$P(A \cap B) = P(A) \cdot P(B) = \frac{18}{36} \cdot \frac{18}{36} = \frac{324}{1296} = \frac{1}{4}$$

43. Each flip of a coin is independent. Let $A = \{\text{heads}\}$, $B = \{\text{heads}\}$, $C = \{\text{tails}\}$ and $D = \{\text{tails}\}$.
$$P(A \cap B \cap C \cap D) = P(A) \cdot P(B) \cdot P(C) \cdot P(D) = \frac{1}{2} \cdot \frac{1}{2} \cdot \frac{1}{2} \cdot \frac{1}{2} = \frac{1}{16}$$

45. The rolls of a pair of dice are independent. Let $A = \{\text{sum of at least 10 on the first roll}\}$, $B = \{\text{sum of at least 10 on the second roll}\}$ and $C = \{\text{sum of at least 10 on the third roll}\}$.
$$P(A \cap B \cap C) = P(A) \cdot P(B) \cdot P(C) = \frac{6}{36} \cdot \frac{6}{36} \cdot \frac{6}{36} = \frac{216}{46{,}656} = \frac{1}{216}$$

47. The choices of a card are independent. Let $A = \{\text{ace on the first draw}\}$ and $B = \{\text{ace on the second draw}\}$.
$$P(A \cap B) = P(A) \cdot P(B) = \frac{4}{52} \cdot \frac{4}{52} = \frac{16}{2704} = \frac{1}{169}$$

49. The choices of a card are independent. Let $A = \{\text{spade on the first draw}\}$ and $B = \{\text{diamond on the second draw}\}$.
$$P(A \cap B) = P(A) \cdot P(B) = \frac{13}{52} \cdot \frac{13}{52} = \frac{169}{2704} = \frac{1}{16}$$

51. The choices of a card are independent. Let $A = \{\text{heart on the first draw}\}$ and $B = \{\text{spade on the second draw}\}$.
$$P(A \cap B) = P(A) \cdot P(B) = \frac{13}{52} \cdot \frac{13}{52} = \frac{169}{2704} = \frac{1}{16}$$

53. a. The choices of a card with replacement are independent Let $A = \{\text{diamond on the first draw}\}$, $B = \{\text{diamond on the second draw}\}$ and $C = \{\text{black card on the third draw}\}$.
$$P(A \cap B \cap C) = P(A) \cdot P(B) \cdot P(C) = \frac{13}{52} \cdot \frac{13}{52} \cdot \frac{26}{52} = \frac{4394}{140{,}608} = \frac{1}{32}$$

b. The choices of a card without replacement are not independent. Let $A = \{\text{diamond card on the first draw}\}$, $B = \{\text{diamond card on the second draw}\}$ and $C = \{\text{black card on the third draw}\}$.
$$P(A \text{ followed by } B \text{ followed by } C) = P(A) \cdot P(B \mid A) \cdot P(C \mid A \text{ and } B) = \frac{13}{52} \cdot \frac{12}{51} \cdot \frac{26}{50} = \frac{4056}{132{,}600} = \frac{13}{425}$$

55. a. The choices of a marble with replacement are independent. Let $A = \{\text{red marble chosen}\}$, $B = \{\text{red marble chosen}\}$, and $C = \{\text{green marble chosen}\}$.
$$P(A \cap B \cap C) = P(A) \cdot P(B) \cdot P(C) = \frac{5}{17} \cdot \frac{5}{17} \cdot \frac{4}{17} = \frac{100}{4913}$$

b. The choices of a marble without replacement are not independent. Let $A = \{\text{red marble chosen}\}$, $B = \{\text{red marble chosen}\}$, and $C = \{\text{green marble chosen}\}$.
$$P(A \text{ followed by } B \text{ followed by } C) = P(A) \cdot P(B \mid A) \cdot P(C \mid A \text{ and } B) = \frac{5}{17} \cdot \frac{4}{16} \cdot \frac{4}{15} = \frac{80}{4080} = \frac{1}{51}$$

57. Let D be the event that a person has been using the drug and let T be the event that the test is positive. The probability we wish to determine is $P(D \mid T)$.
$$P(D \mid T) = \frac{P(D \text{ and } T)}{P(T)}$$
A tree diagram will help us compute the needed probabilities.

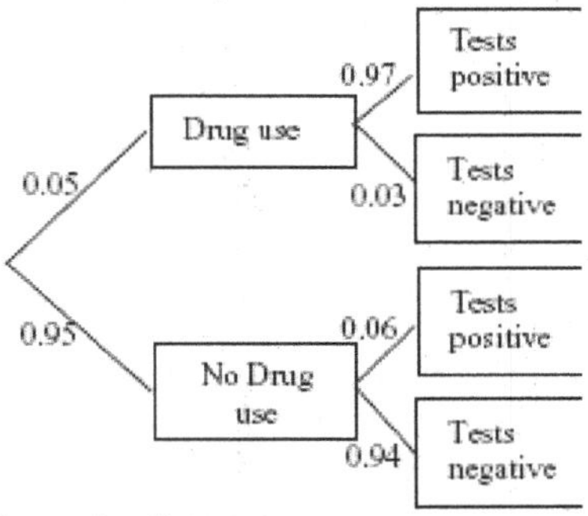

From the diagram,
$P(D \text{ and } T) = P(D) \cdot P(T \mid D) = (0.5)(0.97)$.
To compute $P(T)$ we need to combine two branches from the diagram, one corresponding to a correct positive test result when the person has drug use, and one corresponding to a false positive result when the person has no drug use:
$P(T) = (0.05)(0.97) + (0.95)(0.06)$.
$$P(D \mid T) = \frac{P(D \text{ and } T)}{P(T)} = \frac{(0.05)(0.97)}{(0.05)(0.97) + (0.95)(0.06)} \approx 0.46$$

59. Let D be the event that a person has the disease and let T be the event that the test is positive. The probability we wish to determine is $P(D \mid T)$.

$$P(D \mid T) = \frac{P(D \text{ and } T)}{P(T)}$$

A tree diagram will help us compute the needed probabilities.

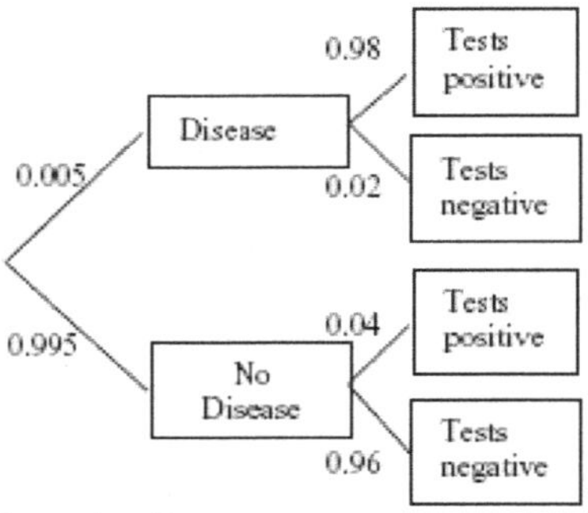

From the diagram,
$P(D \text{ and } T) = P(D) \cdot P(T \mid D) = (0.005)(0.98)$.
To compute $P(T)$ we need to combine two branches from the diagram, one corresponding to a correct positive test result when the person has the disease, and one corresponding to a false positive result when the person does not have the disease:
$P(T) = (0.005)(0.98) + (0.995)(0.04)$.

$$P(D \mid T) = \frac{P(D \text{ and } T)}{P(T)} = \frac{(0.005)(0.98)}{(0.005)(0.98)+(0.995)(0.04)} \approx 0.11$$

THINK ABOUT IT, SECTION 7.6

1. Yes. All casino games have a negative expectation for the player.

2. A person who plays this game many times can expect to lose an average of \$2 each time the game is played.

3. \$9.60, twice the expectation of the \$5 wager.

SECTION EXERCISES 7.6

1. Expectation
$= P(S_1)\cdot S_1 + P(S_2)\cdot S_2 + P(S_3)\cdot S_3 + P(S_4)\cdot S_4 + P(S_5)\cdot S_5$
$= 30\cdot 0.15 + 40\cdot 0.2 + 50\cdot 0.4 + 60\cdot 0.05 + 70\cdot 0.2$
$= 49.5$

3. Let S_1 be the event that the roulette ball lands on a black number, in which case the player wins \$1. There are 38 possible numbers, so $P(S_1) = \frac{18}{38}$. Let S_2 be the event that a black number does not come up, in which case the player loses \$1. Then $P(S_2) = 1 - \frac{18}{38} = \frac{20}{38}$.
$$\text{Expectation} = P(S_1)\cdot S_1 + P(S_2)\cdot S_2 = \frac{18}{38}(1) + \frac{20}{38}(-1) \approx -0.05$$
Expectation is –5 cents.

5. Let S_1 be the event that the wheel stops on Joker (\$40), in which case the player wins \$40. There are 54 possible numbers, so $P(S_1) = \frac{1}{54}$. Let S_2 be the event that the wheel does not stop on Joker, in which case the player loses \$1. Then $P(S_2) = 1 - \frac{1}{54} = \frac{53}{54}$.
$$\text{Expectation} = P(S_1)\cdot S_1 + P(S_2)\cdot S_2 = \frac{1}{54}(40) + \frac{53}{54}(-1) \approx -0.24$$
Expectation is –24 cents.

7. Let S_1 be the event that the wheel stops on \$5, in which case the player wins \$5. There are 54 possible numbers, so $P(S_1) = \frac{7}{54}$. Let S_2 be the event that the wheel does not stop on \$5, in which case the player loses \$1. Then $P(S_2) = 1 - \frac{7}{54} = \frac{47}{54}$.
$$\text{Expectation} = P(S_1)\cdot S_1 + P(S_2)\cdot S_2 = \frac{7}{54}(5) + \frac{47}{54}(-1) \approx -0.22$$
Expectation is –22 cents.

9. Let S_1 be the event that the person dies within one year. Then $P(S_1) = 0.000487$ and the company must pay out \$25,000. Since the company charged \$75 for the policy, the company's actual loss is \$24,925. Let S_2 be the event that the policy holder does not die during the year of the policy. Then $P(S_2) = 0.999513$ and the company keeps the premium of \$75.
Expectation
$$= P(S_1)\cdot S_1 + P(S_2)\cdot S_2 = 0.000487(-24{,}925) + 0.999513(75) = 62.83$$
Expectation is \$62.83.

11. Let S_1 be the event that the person dies within one year. Then $P(S_1) = 0.070471$ and the company must pay out \$10,000. Since the company charged \$495 for the policy, the company' s actual loss is \$9505. Let *S2* be the event that the policy holder does not die during the year of the policy. Then $P(S_2) = 0.929529$ and the company keeps the premium of \$495.

Expectation
$= P(S_1)\cdot S_1 + P(S_2) + S_2$
$= 0.070471(-9505) + 0.929529(495)$
$= -209.71$
Expectation is –\$209.71.

13. Let S_1 be the event that the person dies within one year. Then $P(S_1) = 0.001406$ and the company must pay out \$30,000. If the company charges \$$x$ for the policy, the company's actual loss is \$$(30{,}000 - x)$. Let S_2 be the event that the policyholder does not die during the year of the policy. Then $P(S_2) = 0.998594$ and the company keeps the premium of \$$x$.
Expectation $= P(S_1)\cdot S_1 + P(S_2)\cdot S_2$
$0 = 0.001406(-(30{,}000 - x)) + 0.998594x$
$0 = -42.18 + 0.001406x + 0.998594x$
$42.18 = x$
The company should charge more than \$42.18.

15. Expectation
$= 0.10(100{,}000) + 0.40(60{,}000) + 0.25(30{,}000) + 0.15(0) + 0.08(-20{,}000) + 0.02(-40{,}000)$
$= 10{,}000 + 24{,}000 + 7500 + 0 - 1600 - 800$
$= 39{,}100$
The expected profit is \$39,100.

17. Expectation
$= 0.05(40{,}000) + 0.2(30{,}000) + 0.5(20{,}000) + 0.2(10{,}000) + 0.05(5{,}000)$
$= 2000 + 6000 + 10{,}000 + 2000 + 250$
$= 20{,}250$
The expected profit is \$20,250.

CHAPTER 7 REVIEW EXERCISES

1. {11, 12, 13, 21, 22, 23, 31, 32, 33}

2. {26, 28, 62, 68, 82, 86}

3.

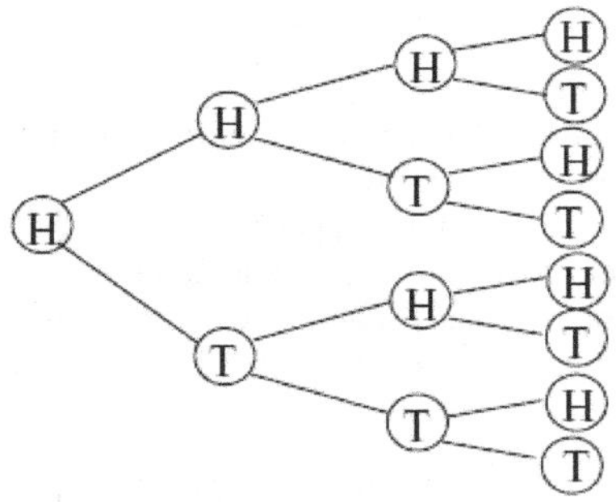

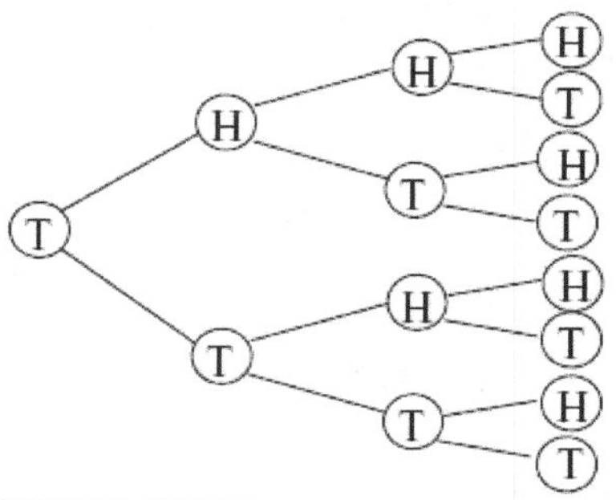

{HHHH, HHHT, HHTH, HHTT, HTHH, HTHT, HTTH, HTTT, THHH, THHT, THTH, THTT, TTHH, TTHT, TTTH, TTTT}

4.

	7	**8**	**9**
A	7A	8A	9A
B	7B	8B	9B

{7A, 8A, 9A, 7B, 8B, 9B}

5. There are 3 choices for style, so $n_1 = 3$. There are 4 choices for color, so $n_2 = 4$. There are 6 choices for sizes, so $n_3 = 6$. By the counting principle, there are $3\cdot4\cdot6 = 72$ shoes available.

6. There are 10 numbers that can be used in each position, so n_1, n_2, n_3 and $n_4 = 10$. By the counting principle, there are $10\cdot10\cdot10\cdot10 = 10{,}000$ different lock combinations possible.

7. There are 2 letters that can be used in the first position, so $n_1 = 2$. There are 12 letters that can be used in the second position, so $n_2 = 12$. There are 10 digits that can be used in the third and fourth positions, so $n_3 = 10$ and $n_4 = 10$. By the counting principle, there are $2\cdot12\cdot10\cdot10 = 2400$ different four-character sequences possible.

8. There are 7 binary digits, the first two of which are required to be different. There are two choices for the first digit, $n_1 = 2$, and only one choice for the second, $n_2 = 1$, since it must be different from the first. There are two choices for each of the five remaining digits, $n_3 = n_4 = n_5 = n_6 = n_7 = 2$. The counting principle then says there are $2\cdot1\cdot2\cdot2\cdot2\cdot2\cdot2 = 64$ biquinary numbers.

9. $7! = 7\cdot6\cdot5\cdot4\cdot3\cdot2\cdot1 = 5040$

10. Evaluating:
$8! - 4! = (8\cdot7\cdot6\cdot5\cdot4\cdot3\cdot2\cdot1) - (4\cdot3\cdot2\cdot1)$
$= 40{,}320 - 24 = 40{,}296$

11. $\dfrac{9!}{2!3!4!}=\dfrac{9\cdot 8\cdot 7\cdot 6\cdot 5\cdot 4!}{2\cdot 1\cdot 3\cdot 2\cdot 1\cdot 4!}=1260$

12. Evaluating:
$$\begin{aligned}P(10,\ 6)&=\frac{10!}{(10-6)!}=\frac{10!}{4!}\\&=\frac{10\cdot 9\cdot 8\cdot 7\cdot 6\cdot 5\cdot 4!}{4!}\\&=10\cdot 9\cdot 8\cdot 7\cdot 6\cdot 5=151,200\end{aligned}$$

13. Evaluating:
$$\begin{aligned}P(8,\ 3)&=\frac{8!}{(8-3)!}=\frac{8!}{5!}=\frac{8\cdot 7\cdot 6\cdot 5!}{5!}\\&=8\cdot 7\cdot 6=336\end{aligned}$$

14. Evaluating:
$$\begin{aligned}\frac{C(6,\ 2)\cdot C(8,\ 3)}{C(14,\ 5)}&=\frac{\frac{6!}{2!(6-2)!}\cdot\frac{8!}{3!(8-3)!}}{\frac{14!}{5!(14-5)!}}\\&=\frac{\frac{6!}{2!4!}\cdot\frac{8!}{3!5!}}{\frac{14!}{5!9!}}\\&=\frac{\frac{6\cdot 5\cdot 4!}{2!4!}\cdot\frac{8\cdot 7\cdot 6\cdot 5!}{3!5!}}{\frac{14\cdot 13\cdot 12\cdot 11\cdot 10\cdot 9!}{5!9!}}\\&=\frac{\frac{6\cdot 5}{2\cdot 1}\cdot\frac{8\cdot 7\cdot 6}{3\cdot 2\cdot 1}}{\frac{14\cdot 13\cdot 12\cdot 11\cdot 10}{5\cdot 4\cdot 3\cdot 2\cdot 1}}\\&=\frac{15\cdot 56}{2002}=\frac{840}{2002}=\frac{60}{143}\end{aligned}$$

15. The order in which the people arrange themselves is important, so the number of ways to receive service is
$$\begin{aligned}P(7,\ 7)&=\frac{7!}{(7-7)!}=\frac{7!}{0!}=\frac{7\cdot 6\cdot 5\cdot 4\cdot 3\cdot 2\cdot 1}{1}\\&=7\cdot 6\cdot 5\cdot 4\cdot 3\cdot 2\cdot 1=5040\end{aligned}$$

16. The order in which the answers are chosen is important, so the number of matches possible is
$$\begin{aligned}P(7,\ 7)&=\frac{7!}{(7-7)!}=\frac{7!}{0!}=\frac{7\cdot 6\cdot 5\cdot 4\cdot 3\cdot 2\cdot 1}{1}\\&=7\cdot 6\cdot 5\cdot 4\cdot 3\cdot 2\cdot 1=5040\end{aligned}$$

17. The order in which the answers are chosen is important, so the number of matches possible is
$$\begin{aligned}P(7,\ 5)&=\frac{7!}{(7-5)!}=\frac{7!}{2!}=\frac{7\cdot 6\cdot 5\cdot 4\cdot 3\cdot 2\cdot 1}{2\cdot 1}\\&=2520\end{aligned}$$

18. We are looking for the number of permutations of the letters *letter*. With $n=6$ (number of letters), $k_1=1$ (number of l's), $k_2=2$ (number of e's), $k_3=2$ (number of t's) and $k_4=1$ (number of r's), we have
$$\frac{6!}{1!2!2!1!}=180$$
There are 180 different arrangements possible.

19. The order in which the heads and tails are chosen is not important, so the number of distinct arrangements is
$$\begin{aligned}C(12,\ 4)&=\frac{12!}{4!(12-4)!}=\frac{12!}{4!8!}\\&=\frac{12\cdot 11\cdot 10\cdot 9\cdot 8!}{4!8!}\\&=\frac{12\cdot 11\cdot 10\cdot 9}{4\cdot 3\cdot 2\cdot 1}=495\end{aligned}$$

20. Because each shift is different, the order of the assignments to shifts is important, so the number of different assignments possible is
$$\begin{aligned}P(5,\ 3)&=\frac{5!}{(5-3)!}=\frac{5!}{2!}=\frac{5\cdot 4\cdot 3\cdot 2!}{2!}\\&=5\cdot 4\cdot 3=60.\end{aligned}$$

21. The order in which the selections are made is not important, so the number of sets that can be chosen is
$$\begin{aligned}C(25,\ 10)&=\frac{25!}{10!(25-10)!}=\frac{25!}{10!15!}\\&=\frac{25\cdot 24\cdot 23\cdot 22\cdot 21\cdot 20\cdot 19\cdot 18\cdot 17\cdot 16\cdot 15!}{10!15!}\\&=\frac{25\cdot 24\cdot 23\cdot 22\cdot 21\cdot 20\cdot 19\cdot 18\cdot 17\cdot 16}{10\cdot 9\cdot 8\cdot 7\cdot 6\cdot 5\cdot 4\cdot 3\cdot 2\cdot 1}\\&=3,268,760.\end{aligned}$$

22. The order in which the stocks are selected is not important, so the number of different portfolios is
$$\begin{aligned}C(11,\ 3)&=\frac{11!}{3!(11-3)!}=\frac{11!}{3!8!}\\&=\frac{11\cdot 10\cdot 9\cdot 8!}{3!8!}\\&=\frac{11\cdot 10\cdot 9}{3\cdot 2\cdot 1}=165.\end{aligned}$$

23. The order of the choices is not important, so the number of possible ways to choose the nondefective monitors, is $C(12,\ 3)$ and the number of ways to choose the defective monitors is $C(3,2)$. Therefore, by the counting principle, there are $C(12,3)\cdot C(3,2)$ different sets.
$$\begin{aligned}C(12,\ 3)\cdot C(3,\ 2)&=\frac{12!}{3!(12-3)!}\cdot\frac{3!}{2!(3-2)!}\\&=\frac{12!}{3!9!}\cdot\frac{3!}{2!1!}\\&=\frac{12\cdot 11\cdot 10\cdot 9!}{3!9!}\cdot\frac{3\cdot 2!}{2!1!}\\&=\frac{12\cdot 11\cdot 10}{3\cdot 2\cdot 1}\cdot\frac{3}{1}\\&=220\cdot 3=660\end{aligned}$$

24. Call the two people that refuse to sit next to each other A and B. Consider two cases. In the first case, A sits at an end of the row. There are 2 ways A can do this. Once A is seated, there are 7 places where B could sit and not be next to A. Then there are 7! ways the remaining people can be seated, for a total of 2·7·7! ways to seat the group if A sits at an end of the row. In the second case, A does not sit at the end of the row. There are 7 ways A can do this. Once A is seated, there are 6 places where B could sit and not be next to A. Then there are 7! ways the remaining people can be seated, for a total of 7·6·7!. Summing the totals from case A and case B gives 282,240 ways.

25. There are $C(4, 4)$ ways of choosing four aces from four aces, $C(4, 4)$ ways of choosing four kings from four kings, and $C(4, 4)$ ways of choosing each of the other possible cards of which there are 13 total. There are 48 choices for the fifth card in the hand. By the counting principle, the number of hands containing four of a kind is

$$\begin{aligned} 13\cdot C(4,\ 4)\cdot 48 &= 13\cdot\frac{4!}{4!(4-4)!}\cdot 48 \\ &= 13\cdot\frac{4!}{4!0!}\cdot 48 = 13\cdot\frac{1}{1}\cdot 48 \\ &= 624. \end{aligned}$$

26. The sample space is {BBBB, BBBG, BBGB, BBGG, BGBB, BGBG, BGGB, BGGG, GBBB, GBBG, GBGB, GBGG, GGBB, GGBG, GGGB, GGGG}. The elements in the event that a family will have one boys and three girls is {BGGG, GBGG, GGBG, GGGB}. Then the probability is

$$P(E) = \frac{n(E)}{n(S)} = \frac{4}{16} = \frac{1}{4}.$$

27. The sample space is {HHH, HHT, HTH, HTT, THH, THT, TTH, TTT}. The elements in the event of getting one head and two tails is {HTT, THT, TTH}. Then the probability is

$$P(E) = \frac{n(E)}{n(S)} = \frac{3}{8}.$$

28. Let E be the event that the employee is a woman.

$$P(E) = \frac{7920}{5739+7290} = \frac{7290}{13{,}029} \approx 0.56$$

29. The total number of students is 2495. Let E be the event that the student is a junior or senior. There are 483 + 445 = 928 students that are juniors or seniors.

$$P(E) = \frac{928}{2495} \approx 0.37$$

30. The total number of students in 2495. Let E be the event that the student is not a graduate student. There are 642 + 549 + 483 + 445 = 2119 students that are not graduate students.

$$P(E) = \frac{2119}{2495} \approx 0.85$$

31. The sample space for tossing two six-sided dice was found in the text. The dice must be considered as distinct, so there are 36 possible outcomes. The elements in the event that the sum of the pips on the upward faces is 9 are {3 6, 4 5, 5 4, 6 3}. Then the probability is

$$P(E) = \frac{n(E)}{n(S)} = \frac{4}{36} = \frac{1}{9}.$$

32. If E is the event of tossing an 11, then E^C is the event of not tossing an 11, and

$$P(E^C) = 1 - P(E) = 1 - \frac{2}{36} = \frac{34}{36} = \frac{17}{18}.$$

33. If E is the event of tossing a sum less than 10, then E^C is the event of tossing a sum of at least ten, and

$$P(E^C) = 1 - P(E) = 1 - \frac{30}{36} = \frac{6}{36} = \frac{1}{6}.$$

34. Let A be the event of rolling an even number, and let B be the event of rolling a number less than 5. From the figure in the text, there are 36 elements in the sample space. $n(A) = 18$, $n(B) = 6$ and $n(A \text{ and } B) = 4$.

$$\begin{aligned} P(A \text{ or } B) &= P(A) + P(B) - P(A \text{ and } B) \\ &= \frac{18}{36} + \frac{6}{36} - \frac{4}{36} = \frac{20}{36} = \frac{5}{9} \end{aligned}$$

35. Let B = {the sum is 9} and A = {the sum is odd}. From the table in the text, there are four possible rolls of the dice for which the sum is 9 and the sum is odd. So $P(A \cap B) = \frac{4}{36} = \frac{1}{9}$.

There are 18 possibilities for which the sum is odd, so $P(A) = \frac{18}{36} = \frac{1}{2}$. We have

$$P(B \mid A) = \frac{P(A \cap B)}{P(A)} = \frac{\frac{1}{9}}{\frac{1}{2}} = \frac{2}{9}.$$

36. Let B = {the sum is 8} and A = {doubles}. From the table in the text, there is one possible roll of the dice for which the sum is 8 and doubles.

So $P(A \cap B) = \frac{1}{36}$. There are 6 possibilities for which the roll is doubles, so $P(A) = \frac{6}{36} = \frac{1}{6}$.

We have

$$P(B \mid A) = \frac{P(A \cap B)}{P(A)} = \frac{\frac{1}{36}}{\frac{1}{6}} = \frac{1}{6}.$$

37. Let A be the event of drawing a heart, and let B be the event of drawing a black card. The sample space consists of 52 cards. $n(A) = 13$, $n(B) = 26$ and $n(A \text{ and } B) = 0$.

$$P(A \text{ or } B) = P(A) + P(B)$$
$$= \frac{13}{52} + \frac{26}{52} = \frac{39}{52} = \frac{3}{4}$$

38. Let A be the event of drawing a heart, and let B be the event of drawing a jack. The sample space consists of 52 cards. $n(A) = 13$, $n(B) = 4$ and $n(A \text{ and } B) = 1$.

$$P(A \text{ or } B) = P(A) + P(B) - P(A \text{ and } B)$$
$$= \frac{13}{52} + \frac{4}{52} - \frac{1}{52} = \frac{16}{52} = \frac{4}{13}$$

39. If E is the event of drawing a three, then E^C is the event of not drawing a three, and

$$P(E^C) = 1 - P(E) = 1 - \frac{4}{52} = \frac{48}{52} = \frac{12}{13}.$$

40. Let B = {the card is red} and A = {the card is not a club}. There are 26 cards that are red and not clubs. So $P(A \cap B) = \frac{26}{52} = \frac{1}{2}$. There are 39 possibilities for cards that are not clubs, so $P(A) = \frac{39}{52} = \frac{3}{4}$. We have

$$P(B \mid A) = \frac{P(A \cap B)}{P(A)} = \frac{\frac{1}{2}}{\frac{3}{4}} = \frac{2}{3}.$$

41. Let E be the event of rolling a sum of 6. Refer to the text. Because there are five ways to roll a sum of 6, there are five favorable outcomes, leaving 31 unfavorable outcomes.

Odds in favor of $E = \frac{5}{31}$

The odds in favor are 5 to 31.

42. Let E be the event of pulling a heart. Because there are thirteen ways to draw a heart, there are thirteen favorable outcomes, leaving 39 unfavorable outcomes.

Odds in favor of $E = \frac{13}{39} = \frac{1}{3}$

The odds in favor are 1 to 3.

43. Because the odds against are 4 to 5, the odds in favor are 5 to 4. Use the formula for calculating the probability of an event when the odds in favor are known.

$$P(E) = \frac{a}{a+b} = \frac{5}{5+4} = \frac{5}{9}$$

44. Make a Punnett square.

Parents	H	h
h	Hh	hh
h	Hh	hh

To have short hair, the rodent must have two h. From the table, two of the four possible genotypes have two h's, so the probability that an offspring will have short hair is $\frac{2}{4} = \frac{1}{2}$.

45. Let A be the event of exactly one card being an ace and let B be the event of exactly one card being a face card. Then,

$$P(A \text{ and } B) = P(A) + P(B) - P(A \text{ or } B)$$
$$= 0.145 + 0.362 - 0.471$$
$$= 0.036$$

46. Let A = {people who like cheese flavored}, and let B = {people who like jalapeno-flavored}. Then, $n(A) = 642$, $n(B) = 487$ and $n(A \text{ and } B) = 302$. The total number of people represented is 1000.

$$P(A \text{ or } B) = P(A) + P(B) - P(A \text{ and } B)$$
$$= \frac{642}{1000} + \frac{487}{1000} - \frac{302}{1000} = \frac{827}{1000}$$

If E is the event of a person liking one of the two flavors, then E^C is the event of not liking either of the two flavors, and

$$P(E^C) = 1 - P(E) = 1 - \frac{827}{1000} = \frac{173}{1000}.$$

47. Because there are 24 chips in the box, there are 24 items in the sample space. Let E be the event of choosing a yellow or white chip. There are 7 yellow chips and 8 white chips, so there are 7 + 8 = 15 items in E. Then the probability is

$$P(E) = \frac{n(E)}{n(S)} = \frac{15}{24} = \frac{5}{8}.$$

48. Let E be the event of getting a red chip. Because there are 4 red chips, there are 4 favorable outcomes, leaving 20 unfavorable outcomes.

Odds in favor of $E = \frac{4}{20} = \frac{1}{5}$

The odds in favor are 1 to 5.

49. Let B = {the chip is yellow} and A = {the chip is not white}. There are seven chips that are yellow and not white. So $P(A\cap B)=\frac{7}{24}$.
There are 16 chips that are not white, so
$P(A)=\frac{16}{24}=\frac{2}{3}$. We have
$$P(B\mid A)=\frac{P(A\cap B)}{P(A)}=\frac{\frac{7}{24}}{\frac{2}{3}}=\frac{7}{16}.$$

50. Let E = {no red chips are drawn}. There are 24 possibilities for the first chip drawn, 23 possibilities for the second chip drawn, 22 for the third, 21 for the fourth and 20 for the fifth. Thus, $n(S) = 24{\cdot}23{\cdot}22{\cdot}21{\cdot}20 = 5{,}100{,}480$. The first time a chip is drawn, there are 20 chips that are not red, the second time there are 19 chips that are not red, 18 chips the third time 17 chips the fourth time, and 16 chips the fifth time, so $n(E) = 20{\cdot}19{\cdot}18{\cdot}17{\cdot}16 = 1{,}860{,}480$.
$$P(E)=\frac{1{,}860{,}480}{5{,}100{,}480}=\frac{646}{1771}$$

51. Let A = {a yellow is selected first}, B = {a white is selected second} and C = {a yellow is selected third}. Then
$$P(A \text{ followed by } B \text{ followed by } C)$$
$$=P(A)\cdot P(B\mid A)\cdot P(C\mid A \text{ and } B)$$
$$=\frac{7}{24}\cdot\frac{8}{23}\cdot\frac{6}{22}=\frac{336}{12{,}144}=\frac{7}{253}$$

52. $n(S) = 23{,}500$. Let E be the event that the person voted against the proposition. Then
$$P(E)=\frac{2527+5370+712}{23{,}500}=\frac{8609}{23{,}500}\approx 0.37.$$

53. $n(S) = 23{,}500$. Let A = {Democrats}, and let B = {Independents}. Then, from the table, $n(A) = 8452 + 2527 + 894 = 11{,}873$, $n(B) = 1225 + 712 + 686 = 2623$ and $n(A \text{ and } B) = 0$.
$$P(A \text{ or } B)=P(A)+P(B)-P(A \text{ and } B)$$
$$=\frac{11{,}873}{23{,}500}+\frac{2623}{23{,}500}-0$$
$$=\frac{14{,}496}{23{,}500}\approx 0.62$$

54. $n(S) = 23{,}500$. Let E be the event that the person abstained from voting and is not a Republican. Then
$$P(E)=\frac{894+686}{23{,}500}=\frac{1580}{23{,}500}\approx 0.07.$$

55. $n(S) = 23{,}500$. Let B = {voted for the proposition} and A = {registered Independent}. From the table, $n(A\cap B) = 1225$, $n(A) = 1225 + 712 + 686 = 2623$, and $n(S) = 23{,}500$. We have
$$P(B\mid A)=\frac{P(A\cap B)}{P(A)}=\frac{\frac{1225}{23{,}500}}{\frac{2623}{23{,}500}}\approx 0.47.$$

56. $n(S) = 23{,}500$. Let B = {registered Democrat} and A = {voted against the proposition}. From the table, $n(A\cap B) = 2527$ and $n(A) = 2527 + 5370 + 712 = 8609$. We have
$$P(B\mid A)=\frac{P(A\cap B)}{P(A)}=\frac{\frac{2327}{23{,}500}}{\frac{8609}{23{,}500}}\approx 0.29.$$

57. The rolls of a pair of dice are independent. Let A = {6 on the first roll}, B = {6 on the second roll} and C = {6 on the third roll}. Then
$$P(A\cap B\cap C)=P(A)\cdot P(B)\cdot P(C)$$
$$=\frac{1}{6}\cdot\frac{1}{6}\cdot\frac{1}{6}=\frac{1}{216}$$

58. Let E = {at least one 6}; then E^C = {no 6's}. Because on each toss of the die there are six possible outcomes, $n(S) = 6{\cdot}6{\cdot}6{\cdot}6{\cdot}6 = 7776$. On each toss of the die there are five numbers that are not 6's. Therefore, $n(E^C) = 5{\cdot}5{\cdot}5{\cdot}5{\cdot}5 = 3125$.
$$P(E)=1-P(E^C)=1-\frac{3125}{7776}=\frac{4651}{7776}\approx 0.60$$

59. Let E = {exactly two 6's}. Because on each toss of the die there are six possible outcomes, $n(S) = 6{\cdot}6{\cdot}6{\cdot}6{\cdot}6 = 7776$. On each toss of the die there are five numbers that are not 6's and one number that is a 6. The number of possible arrangements of the two sixes is $C(5, 2) = 10$. Therefore, $n(E) = 10{\cdot}5{\cdot}5{\cdot}5{\cdot}1{\cdot}1 = 1250$.
$$P(E)=\frac{1250}{7776}\approx 0.16$$

60. Let E = {at least one spade is drawn}; then E^C = {no spade is drawn}. There are 52 possibilities for each card drawn. Thus, $n(S) = 52{\cdot}52{\cdot}52{\cdot}52 = 7{,}311{,}616$. Each time a card is drawn, there are 39 cards that are not spades, so $n(E^C) = 39{\cdot}39{\cdot}39{\cdot}39 = 2{,}313{,}441$.
$$P(E)=1-P(E^C)=1-\frac{2{,}313{,}441}{7{,}311{,}616}$$
$$=\frac{4{,}998{,}175}{7{,}311{,}616}=\frac{175}{256}$$

61. Let D be the event that a dog has the disease and let T be the event that the test is positive. The probability we wish to determine is $P(D/T)$. Then

$$P(D \mid T) = \frac{P(D \text{ and } T)}{P(T)}.$$

A tree diagram will help us compute the needed probabilities.

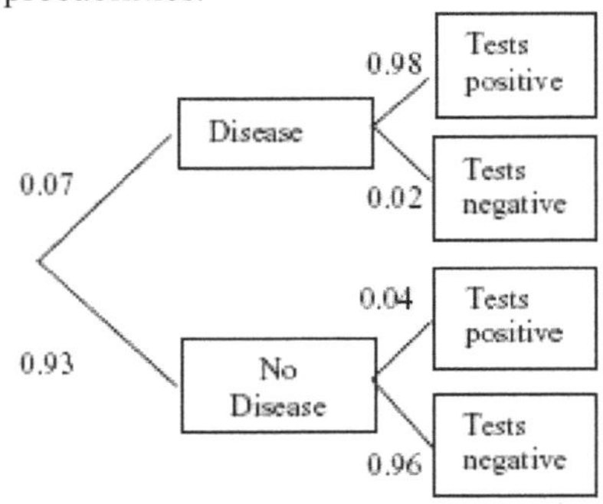

From the diagram, $P(D \text{ and } T) = P(D) \cdot P(T \mid D) = (0.07)(0.98)$. To compute $P(T)$ we need to combine two branches from the diagram, one corresponding to a correct positive test result when the dog has the disease, and one corresponding to a false positive result when the dog does not have the disease:

$P(T) = (0.07)(0.98) + (0.93)(0.04)$. Then

$$P(D \mid T) = \frac{P(D \text{ and } T)}{P(T)} = \frac{(0.07)(0.98)}{(0.07)(0.98) + (0.93)(0.04)} \approx 0.648$$

62. Let A = {rain the first day}, B = {rain the second day}, C = {no rain the third day} and D = {no rain the fourth day}. Then

$$\begin{aligned} &P(A \text{ followed by } B \text{ followed by } C \text{ followed by } D) \\ &= P(A) \cdot P(B \mid A) \cdot P(C \mid A \text{ and } B) \cdot P(D \mid A, B \text{ and } C) \\ &= 0.15 \cdot 0.65 \cdot 0.35 \cdot 0.85 \\ &\approx 0.029 \end{aligned}$$

63. $1 - (0.988)^{12} \approx 0.135$

64. Let S_1 be the event that a 1 is rolled, in which case the player loses \$3. There is 1 possible number, so $P(S_1) = \frac{1}{6}$. Let S_2 be the event that a 2 is rolled, in which case the player loses \$2. $P(S_2) = \frac{1}{6}$. Let S_3 be the event that a 3 is rolled, in which case the player loses \$1. There is 1 possible number, so $P(S_3) = \frac{1}{6}$. Let S_4 be the event that a 4 is rolled, in which case the player wins \$0. $P(S_4) = \frac{1}{6}$. Let S_5 be the event that a 5 is rolled, in which case the player wins \$1. There is 1 possible number, so $P(S_5) = \frac{1}{6}$.

Let S_6 be the event that a 6 is rolled, in which case the player wins \$2. $P(S_6) = \frac{1}{6}$.

$$\begin{aligned} &\text{Expectation} \\ &= P(S_1) \cdot S_1 + P(S_2) \cdot S_2 + P(S_3) \cdot S_3 \\ &\quad + P(S_4) \cdot S_4 + P(S_5) \cdot S_5 + P(S_6) \cdot S_6 \\ &= \frac{1}{6}(-3) + \frac{1}{6}(-2) + \frac{1}{6}(-1) \\ &\quad + \frac{1}{6}(0) + \frac{1}{6}(1) + \frac{1}{6}(2) \\ &= -\frac{1}{2} \end{aligned}$$

Expectation is –50 cents.

65. Let S_1 be the event that both coins are tails, in which case the player wins \$5. There is 1 outcome in the sample space of 4, so $P(S_1) = \frac{1}{4}$.

Let S_2 be the event that one coin is heads and one is tails, in which case the player loses \$2. There are two outcomes in the sample space of 4, so $P(S_2) = \frac{2}{4} = \frac{1}{2}$. Let S_3 be the event that both coins are heads, in which case the player neither wins nor loses money. There is 1 outcome in the sample space of 4, so $P(S_3) = \frac{1}{4}$.

$$\begin{aligned} &\text{Expectation} \\ &= P(S_1) \cdot S_1 + P(S_2) \cdot S_2 + P(S_3) \cdot S_3 \\ &= \frac{1}{4}(5) + \frac{1}{2}(-2) + \frac{1}{4}(0) \\ &= 0.25 \end{aligned}$$

Expectation is 25 cents.

66. Let S_1 be the event that the first prize is won, in which case the player wins \$200. $P(S_1) = \frac{1}{800}$.

Let S_2 be the event that the second prize is won, in which case the player wins \$75.

$P(S_2) = \frac{4}{800}$. It costs \$1 to buy a ticket.

$$\begin{aligned} &\text{Expectation} \\ &= P(S_1) \cdot S_1 + P(S_2) \cdot S_2 - 1 \\ &= \frac{1}{800}(200) + \frac{4}{800}(75) - 1 \\ &\approx -0.37 \end{aligned}$$

Expectation is –37 cents.

67. The sample space for tossing two six-sided dice is found in the text. There are 36 possible outcomes. There are 3 elements in the event that the sum of the pips on the upward faces is 4. Then the probability is

$$P(E)=\frac{n(E)}{n(S)}=\frac{3}{36}=\frac{1}{12}.$$

Multiply the probability by 65 and find that we can expect a sum of 4 about 5.4 times.

68. Let S_1 be the event that the person dies within one year. Then $P(S_1) = 0.000595$ and the company must pay out \$40,000. Since the company charged \$320 for the policy, the company's actual loss is \$39,680. Let S_2 be the event that the policy holder does not die during the year of the policy. Then $P(S_2) = 0.999405$ and the company keeps the premium of \$320.

Expectation
$= P(S_1)\cdot S_1 + P(S_2)\cdot S_2$
$= 0.000595(-39{,}680) + 0.999405(320)$
$= 296.20$
Expectation is \$296.20 .

69. Let S_1 be the event that the person dies within one year. Then $P(S_1) = 0.001188$ and the company must pay out \$25,000. Since the company charged \$795 for the policy, the company's actual loss is \$24,205. Let S_2 be the event that the policy holder does not die during the year of the policy. Then $P(S_2) = 0.998812$ and the company keeps the premium of \$795.

Expectation
$= P(S_1)\cdot S_1 + P(S_2)\cdot S_2$
$= 0.001188(-24{,}205) + 0.998812(795)$
$= 765.30$
Expectation is \$765.30.

70. Expectation
$= 0.20(25{,}000) + 0.25(15{,}000) + 0.20(10{,}000) + 0.15(5{,}000) + 0.10(0) + 0.10(-5000)$
$= 5000 + 3750 + 2000 + 750 + 0 - 500$
$= 11{,}000$
The expected profit is \$11,000.

Chapter 8: Statistics

THINK ABOUT IT, SECTION 8.1

1. Descriptive statistics involves the collection, organization, summarization, and presentation of data. Inferential statistics interprets and draws conclusions from the data.

2. The mean will change because 2700 replaces 2500 as part of a sum. The median will not change because 2700 does affect the number of scores above the median.

3. Yes. The mode is the most frequently occurring data point and therefore is in the data set.

4. The median is the data value in the middle of an ordered list of numbers, so the median of this data set is 3. The mean is the sum of the data values divided by the number of data values. Since the data value of 20 is quite large, the sum of the data values will be quite large. So, one can expect the mean to be greater than the median in this instance.

5. A weighted mean assigns more importance (weights) to some values over others. For instance, the calculation of GPA (grade point average) assigns different weights to different grades.

SECTION EXERCISES 8.1

1. $\bar{x}=\dfrac{\sum x}{n}=\dfrac{2+7+5+7+14}{5}=7$

   Ranking the numbers from smallest to largest gives 2, 5, 7, 7, 14. The middle number is 7.Thus 7 is the median.
   The number 7 occurs more often than other numbers. Thus 7 is the mode.

3. $\bar{x}=\dfrac{11+8+2+5+17+39+52+42}{8}=22$

   Ranking the numbers from smallest to largest gives 2, 5, 8, 11, 17, 39, 42, 52. The two middle numbers are 11 and 17. The mean of 11 and 17 is 14. Thus 14 is the median. No number occurs more often than the others, so there is no mode.

5. $\bar{x}=\dfrac{\sum x}{n}=\dfrac{2.1+4.6+8.2+3.4+5.6+8.0+9.4+12.2+56.1+78.2}{10}\approx 18.8$

   Ranking the numbers from smallest to largest gives 2.1, 3.4, 4.6, 5.6, 8.0, 8.2, 9.4, 12.2, 56.1, 78.2. The two middle numbers are 8.0 and 8.2. The mean of 8.0 and 8.2 is 8.1. Thus 8.1 is the median. No number occurs more often than the others, so there is no mode.

7. $\bar{x}=\dfrac{\sum x}{n}=\dfrac{255+178+192+145+202+188+178+201}{8}\approx 192.4$

   Ranking the numbers from smallest to largest gives 145, 178, 178, 188, 192, 201, 202, 255. The two middle numbers are 188 and 192. The mean of 188 and 192 is 190. Thus 190 is the median. The number 178 occurs more often than the other numbers. Thus 178 is the mode.

9. $\bar{x}=\dfrac{\sum x}{n}=\dfrac{-12+(-8)+(-5)+(-5)+(-3)+0+4+9+21}{9}\approx 0.1$

   Ranking the numbers from smallest to largest gives –12, –8, –5, –5, –3, 0, 4, 9, 21. The middle number is – 3.Thus –3 is the median. The number –5 occurs more often than the other numbers. Thus –5 is the mode.

11. $\bar{x}=\dfrac{\sum x}{n}=\dfrac{1439}{36}\approx 40.0$

    Ranking the numbers from smallest to largest shows the two middle numbers to be 35 and 35. Thus 35 is the median.

13. a. Answers will vary.
    b. Answers will vary.

15. The A is worth 4 points with a weight of 3; the B is worth 3 points with a weight of 4; the C^+ is worth 2.33 with a weight of 3; and the B^- is worth 2.67 with a weight of 2. The sum of all the weights is 3 + 3 + 4 + 3 + 2, or 15.

    Weighted mean $=\dfrac{(4\times 3)+(4\times 3)+(3\times 4)+(2.33\times 3)+(2.67\times 2)}{15}=\dfrac{48.33}{15}\approx 3.22$

    Jerry's GPA for the fall semester is 3.22.

17. Find the weighted mean of both semesters. The sum of all the weights is 46 + 12, or 58.

$$\text{Weighted mean} = \frac{(3.24\times 46)+(3.86\times 12)}{58} = \frac{195.36}{58} \approx 3.37$$

Tessa's cumulative GPA is 3.37.

19. $$\text{Weighted Mean} = \frac{\sum(x\cdot w)}{\sum w}$$

$$= \frac{(70\times 10\%)+(65\times 10\%)+(82\times 10\%)+(94\times 10\%)+(85\times 10\%)+(92\times 20\%)+(80\times 30\%)}{15} = \frac{82}{1} = 82$$

21. $$\text{Players slugging average} = \frac{s+2d+3t+4h}{n} = \frac{73+2(36)+3(9)+4(54)}{458} = \frac{388}{458} \approx 0.847$$

23. $$\text{Players slugging average} = \frac{s+2d+3t+4h}{n} = \frac{94+2(33)+3(1)+4(49)}{535} = \frac{359}{535} \approx 0.671$$

25. $$\text{Mean} = \frac{\sum(x\cdot f)}{\sum f} = \frac{(2\cdot 6)+(4\cdot 5)+(5\cdot 6)+(9\cdot 3)+(10\cdot 1)+(14\cdot 2)+(19\cdot 1)}{24} = \frac{146}{24} \approx 6.1 \text{ points}$$

If the data were written in a list (6 twos, 5 fours, 6 fives, 3 nines, 1 ten, 2 fourteens, 1 nineteen), the middle number would be 5. Thus 5 points is the median. 2 points and 5 points occur most. Thus 2 and 5 points are the mode.

27. $$\text{Mean} = \frac{\sum(x\cdot f)}{\sum f} = \frac{(2\cdot 1)+(4\cdot 2)+(6\cdot 7)+(7\cdot 12)+(8\cdot 10)+(9\cdot 4)+(10\cdot 3)}{39} = \frac{282}{39} \approx 7.2$$

The middle score is 7. Thus 7 is the median. The score of 7 occurs more often than any other score. Thus 7 is the mode.

29. $$\text{Midrange} = \frac{56^\circ + 76^\circ}{2} = 64^\circ$$

31. $$\text{Midrange} = \frac{-56^\circ\text{F} + 44^\circ\text{F}}{2} = -6^\circ\text{F}$$

33. Let x be the score Ruben needs on the next test. The mean score has a weight of 6 and the new score has a weight of 1, so

$$80 = \frac{(78\times 6)+(x\times 1)}{7}$$
$$560 = 468 + x$$
$$92 = x$$

35. a. $\bar{x} = \dfrac{0.336+0.213}{2} \approx 0.275$

b. Batting average

$$\frac{92+60}{274+282} \approx 0.273$$

c. No.

THINK ABOUT IT, SECTION 8.2

1. The range is the difference between the largest and smallest values in the data set.

2. Variance measures how spread out data are from the mean.

3. The standard deviation of a data set is the square root of the variance.

4. If the standard deviation of a data set is zero, all values of the data set are equal.

5. The data set 2, 4, 6, 8, 10 has the larger standard deviation because its values are more spread out from the mean.

6. Small. This would indicate that each box was getting very close to the same amount of cereal.

SECTION EXERCISES 8.2

1. The highest temperature is 63°F and the lowest temperature is –21°F. The range is 63°F – (–21°F)= 84°F.

3. The greatest data value is 22 and the least data value is 1. The range is 22 – 1= 21.

$\bar{x}=\frac{1+2+5+7+8+19+22}{7}=\frac{64}{7}\approx 9.1429$

x	$x-\bar{x}$	$(x-\bar{x})^2$
1	1 – 9.1429	$(-8.1429)^2 = 66.3068$
2	2 – 9.1429	$(-7.1429)^2 = 51.0210$
5	5 – 9.1429	$(-4.1429)^2 = 17.1636$
7	7 –9.1429	$(-2.1429)^2 = 4.5920$
8	8 –9.1429	$(-1.1429)^2 = 1.3062$
19	19 – 9.1429	$(9.8571)^2 = 97.1624$
22	22 – 9.1429	$(12.8571)^2 = 165.3050$
		402.857

$\frac{\text{sum of the (deviations)}^2}{n-1}=\frac{402.857}{7-1}=\frac{402.857}{6}\approx 67.1428$

The square root of the quotient is

$s=\sqrt{67.1428}\approx 8.2$

The variance is the square of the standard deviation. $s^2=(\sqrt{67.1428})^2\approx 67.1$

5. The greatest data value is 4.8 and the least data value is 1.5. The range is 4.8 – 1.5 = 3.3.

$\bar{x}=\frac{2.1+3.0+1.9+1.5+4.8}{5}=\frac{13.3}{5}=2.66$

x	$x-\bar{x}$	$(x-\bar{x})^2$
2.1	2.1 – 2.66	$(-0.56)^2 = 0.3136$
3.0	3.0 – 2.66	$(0.34)^2 = 0.1156$
1.9	1.9 – 2.66	$(-0.76)^2 = 0.5776$
1.5	1.5 – 2.66	$(-1.16)^2 = 1.3456$
4.8	4.8 – 2.66	$(2.14)^2 = 4.5796$
		6.932

$\frac{\text{sum of the (deviations)}^2}{n-1}=\frac{6.932}{5-1}=\frac{6.932}{4}=1.733$

The square root of the quotient is

$s=\sqrt{1.733}\approx 1.3.$

Variance is the square of the standard deviation. Thus the variance is $s^2=(\sqrt{1.733})^2\approx 1.7.$

7. The greatest data value is 100 and the least data value is 48. The range is 100 – 48 = 52.

$\bar{x}=\frac{48+91+87+93+59+68+92+100+81}{9}$

$=\frac{719}{9}=79.889$

x	$x-\bar{x}$	$(x-\bar{x})^2$
48	48 – 79.889	$(-31.889)^2 = 1016.90$
91	91 – 79.889	$(11.111)^2 = 123.454$
87	87 – 79.889	$(7.111)^2 = 50.566$
93	93 – 79.889	$(13.11)^2 = 171.898$
59	59 – 79.889	$(-20.889)^2 = 436.350$
68	68 – 79.889	$(-11.889)^2 = 141.348$
92	92 – 79.889	$(12.111)^2 = 146.676$
100	100 – 79.889	$(20.111)^2 = 404.452$
81	81 – 79.889	$(1.111)^2 = 1.234$
		2492.878

$\frac{\text{sum of the (deviations)}^2}{n-1}=\frac{2492.878}{9-1}=\frac{2492.878}{8}=311.610$

The square root of the quotient is

$s=\sqrt{311.610}\approx 17.7.$

Variance is the square of the standard deviation. Thus the variance is

$s^2=(\sqrt{311.610})^2\approx 311.6.$

9. All of the data values are 4. Thus, the range is 4 – 4 = 0.

$\bar{x}=\frac{4+4+4+4+4+4+4+4+4+4+4+4+4+4+4+4+4}{17}=4$

x	$x-\bar{x}$	$(x-\bar{x})^2$
4	4 – 4	$(0)^2 = 0$
4	4 – 4	$(0)^2 = 0$
4	4 – 4	$(0)^2 = 0$
4	4 – 4	$(0)^2 = 0$
4	4 – 4	$(0)^2 = 0$
4	4 – 4	$(0)^2 = 0$
etc.	etc.	etc.
		0

$\frac{\text{sum of the (deviations)}^2}{n-1}=\frac{0}{14-1}=\frac{0}{13}=0.$

The square root of the quotient is $s=\sqrt{0}=0.$

Variance is the square of the standard deviation. Thus the variance is $s^2=(\sqrt{0})^2=0.$

11. The greatest data value is 11 and the least data value is –12. The range is 11 – (–12) = 23.

$$\bar{x} = \frac{-8+(-5)+(-12)+(-1)+4+7+11}{7} = \frac{-4}{7} \approx -0.571$$

x	$x-\bar{x}$	$(x-\bar{x})^2$
–8	–8 – (–0.571)	$(-7.429)^2 = 55.190$
–5	–5 – (–0.571)	$(-4.429)^2 = 19.616$
–12	–12– (–0.571)	$(-11.429)^2 = 130.622$
–1	–1 – (–0.571)	$(-0.429)^2 = 0.184$
4	4 – (–0.571)	$(4.571)^2 = 20.894$
7	7– (–0.571)	$(7.571)^2 = 57.320$
11	11 – (–0.571)	$(11.571)^2 = 133.888$
		417.714

$$\frac{\text{sum of the (deviations)}^2}{n-1} = \frac{417.714}{7-1} = \frac{417.714}{6} = 69.619$$

The square root of the quotient is $s = \sqrt{69.619} \approx 8.3$.

Variance is the square of the standard deviation. Thus the variance is $s^2 = (\sqrt{69.619})^2 \approx 69.6$.

13. Opinions will vary. However, many climbers would consider rope B to be safer because of its smaller standard deviation in breaking strength.

15. The students in the college statistics course because the range of weights is greater.

17. Using a calculator, the mean, to the nearest hundredth, is $\bar{x} \approx 23.67$ mpg.
The sample standard deviation (given as Sx on a TI-83/84 calculator), to the nearest hundredth, is $s \approx 3.92$ mpg.

19. Using a calculator, the mean, to the nearest hundredth, is $\bar{x} \approx 493.60$ cal.
The sample standard deviation (given as Sx on a TI-83/84 calculator), to the nearest hundredth, is $s \approx 20.30$ cal.

21. Using a calculator, the mean, to the nearest hundredth, is $\bar{x} \approx 4.27$ hr.
The sample standard deviation (given as Sx on a TI-83/84 calculator), to the nearest hundredth, is $s \approx 0.69$ hr.

23. a. Using a calculator, the mean $(\bar{x})$, to the nearest second, is 210 sec, or 3 min 30 sec, and the sample standard deviation (Sx), to the nearest second, is 26 sec.

b. Yes.
Add and Subtract from the mean:
210 + 26 = 236 sec = 3:56 (min:sec)
210 – 26 = 184 sec = 3:04 (min:sec)
Therefore, the song length greater than 3:56 min:sec and lower than 3:04 min:sec are 4:02 min:sec, 3:01 min:sec, and 2:27min:sec.

THINK ABOUT IT, SECTION 8.3

1. The z-score for a given data value x is the number of standard deviations that x is above or below the mean of the data.

2. A z-score of –1 indicates that the value is one standard deviation below the mean.

3. It means that 58% of the data is smaller than 34.

4. Twenty-five percent of the data is less than 84.

5. Fifty percent of the data set lies between Q_1 and Q_2.

SECTION EXERCISES 8.3

1. a. $z_{85} = \frac{85-75}{11.5} \approx 0.87$

b. $z_{95} = \frac{95-75}{11.5} \approx 1.74$

c. $z_{50} = \frac{50-75}{11.5} \approx -2.17$

d. $z_{75} = \frac{75-75}{11.5} = 0$

3. a. $z_{6.2} = \frac{6.2-6.8}{1.9} \approx -0.32$

b. $z_{7.2} = \frac{7.2-6.8}{1.9} \approx 0.21$

c. $z_{9.0} = \frac{9.0-6.8}{1.9} \approx 1.16$

d. $z_{5.0} = \frac{5.0-6.8}{1.9} \approx -0.95$

5. a. $z_{110.5} = \dfrac{110.5-119.4}{13.2} \approx -0.67$

   b. Solving for x:

   $$z_x = \frac{x-119.4}{13.2} = 2.15$$
   $$x - 119.4 = 28.4$$
   $$x = 147.78$$

   Her systolic blood pressure reading was 147.78 mm Hg.

7. a. $z_{214} = \dfrac{214-182}{44.2} \approx 0.72$

   b. Solving for x:

   $$z_x = \frac{x-182}{44.2} = -1.58$$
   $$x - 182 = -69.836$$
   $$x \approx 112.16$$

   His blood cholesterol level is approximately 112.16 mg/dl.

9. a. $z_{65} = \dfrac{65-72}{8.2} \approx -0.85$

   b. $z_{102} = \dfrac{102-130}{18.5} \approx -1.51$

   c. $z_{605} = \dfrac{605-720}{116.4} \approx -0.99$

   The 65 is the highest relative score.

11. Percentile of $455 = \dfrac{4256}{7210} \cdot 100 \approx 59$

    Shaylen's score was at the 59th percentile.

13. Kevin's percentile $= \dfrac{x}{9840} \cdot 100 = 65$

    $$x = \frac{65 \cdot 9840}{100} = 6396$$

    6396 students scored lower than Kevin.

15. a. The median is by definition the 50th percentile. Therefore, 50% of the four-person family incomes are more than \$66,650.

    b. Because \$178,500 is the 90th percentile, 100 – 90 = 10%of the four-person family incomes are more than \$178,500.

    c. 90 – 50 = 40%of the four-person family incomes are between \$66,650 and \$178,500.

17. 1, 3, 4, 5, 5, 7, 8, 10, 10, 10, 12, 15, 18, 26, 28, 32, 41, 85

    The median (or second quartile) is the average of the 9th and 10th data values. $\dfrac{10+10}{2} = 10$

    The first quartile is the median of the 1st through 9th data values. 5.

    The third quartile is the median of the 10th through 18th data values. 26. $Q_1 = 5$, $Q_2 = 10$, $Q_3 = 26$

19. Northeast:
    minimum: 227,400, maximum: 346,000
    $Q_1 = 264{,}300$, $Q_2 = 315{,}800$, $Q_3 = 343{,}600$

    Midwest:
    minimum: 169,700, maximum: 216,900
    $Q_1 = 178{,}000$, $Q_2 = 197{,}600$, $Q_3 = 208{,}600$

    South:
    minimum: 148,000, maximum: 217,700
    $Q_1 = 163{,}400$, $Q_2 = 194{,}800$, $Q_3 = 203{,}700$

    West:
    minimum: 196,400, maximum: 337,700
    $Q_1 = 238{,}500$, $Q_2 = 263{,}700$, $Q_3 = 330{,}900$

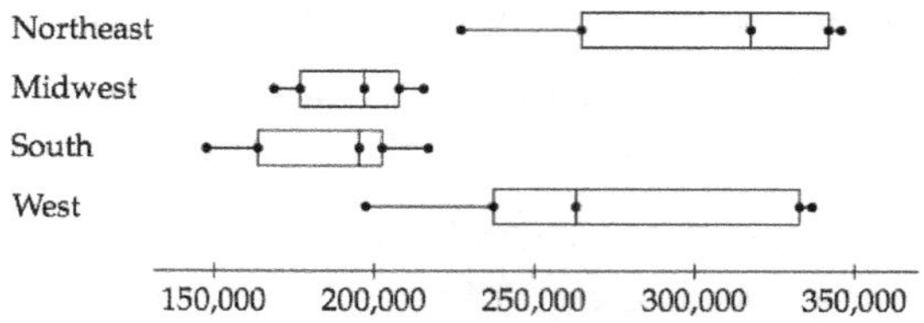

Answers will vary. Here are some possibilities. The region with the lowest median was the South; the region with the highest median was the Northeast. The range of prices was greatest for the West.

21. CBX-21:
    minimum: 44, maximum: 60
    $Q_1 = 46.5$, $Q_2 = 49.5$, $Q_3 = 52.5$

    PHT-34:
    minimum: 38, maximum: 50
    $Q_1 = 44$, $Q_2 = 46$, $Q_3 = 47$

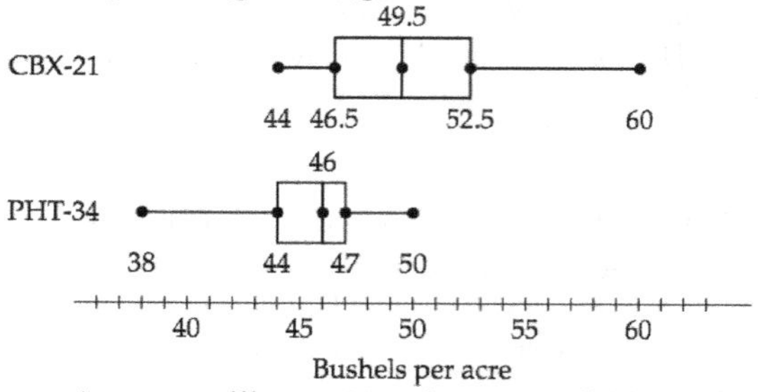

Answers will vary. Here is one possibility: The maximum number of bushels cultivated per acres for PHT-34 is approximately equal to the median number of bushels cultivated per acres for CBX-21.

THINK ABOUT IT, SECTION 8.4

1. A normal distribution with mean 0 and standard deviation 1.

2. For data that are normally distributed, 50% of the data lie below the mean. For data that are normally distributed, 50% of the data lie above the median.

3. According to the Empirical Rule for normally distributed data, approximately 68% of the data lie between plus or minus 1 standard deviation from the mean.

4. If a data value is randomly selected from normally distributed data, the probability that it will be more than 3 standard deviations is approximately 1%.

SECTION EXERCISES 8.4

1. a. The percent of 19th -century boys who were at least 65 inches is 5.8%.

   b. The percent of data in all classes with a lower boundary of 55 inches and an upper boundary of 64 inches is the sum of the percents: 46% + 41% = 87%.
   The probability that a 19th-century boy selected at random will be at least 55 inches tall but less than 65 inches tall is 0.87.

3. a. 95% of the data lies within 2 standard deviations of the mean.

   b. 1 – (50 + 34) = 16%
   16% of the data lie more than 1 standard deviation above the mean.

   c. 34 + 34 + 13.5 = 81.5%
   81.5% of the data lies between 1 standard deviation below the mean and 2 standard deviations above the mean.

5. a. 12 ounces is 2 standard deviations below the mean, and 30 ounces is 1 standard deviation above the mean.
   34 + 34 + 13.5 = 81.5%.
   81.5% of the parcels weighed between 12 and 30 ounces.

   b. 42 ounces is 3 standard deviations above the mean. Therefore, 0.15% of the parcels weighed more than 42 ounces.

7. a. 68 miles per hour is 1 standard deviation above the mean. (0.16)(8000) = 1280.
   1280 vehicles had a speed of more than 68 miles per hour.

   b. 40 miles per hour is 3 standard deviation below the mean. (0.0015)(8000) = 12.
   12 vehicles had a speed of less than 40 miles per hour.

9. When z is 1.5, A is 0.433. When z is 0, A is 0.Therefore, the area of the distribution between $z = 0$ and $z = 1.5$ is 0.433 square unit.

11. The area of the distribution between $z = 0$ and $z = -1.85$ is equal to the area of the distribution between $z = 0$ and $z = 1.85$, due to symmetry. When z is 1.85, A is 0.468. When z is 0, A is 0.Therefore, the area of the distribution between $z = 0$ and $z = -1.85$ is 0.468 square unit.

13. When z is 1.9, A is 0.471. When z is 1, A is 0.341. 0.471 – 0.341 = 0.130.
    The area of the distribution between $z = 1$ and $z = 1.9$ is 0.130 square unit.

15. The area of the distribution between $z = 0$ and $z = -1.47$ is equal to the area of the distribution between $z = 0$ and $z = 1.47$, due to symmetry. When z is 1.47, A is 0.429. When z is 0, A is 0.When z is 1.64, A is 0.449. When z is 0, A is 0.0.429 + 0.449 = 0.878.
    The area of the distribution between $z = -1.47$ and $z = 1.64$ is 0.878 square unit.

17. The area of the distribution between $z = 0$ and $z = 1.3$ is 0.403 square unit. The area of the distribution to the right of $z = 0$ is 0.500 square unit. 0.500 – 0.403 = 0.097.
    The area of the distribution where $z > 1.3$ is 0.097 square unit.

19. The area of the distribution between $z = 0$ and $z = -2.22$ is 0.487 square unit. The area of the distribution to the left of $z = 0$ is 0.500 square unit. 0.500 – 0.487 = 0.013.
    The area of the distribution where $z < -2.22$ is 0.013 square unit.

21. The area of the distribution between $z = 0$ and $z = -1.45$ is 0.426 square unit. The area of the distribution to the right of $z = 0$ is 0.500 square unit. 0.426 + 0.500 = 0.926.
    The area of the distribution where $z > -1.45$ is 0.926 square unit.

23. The area of the distribution between $z = 0$ and $z = 2.71$ is 0.497 square unit. The area of the distribution to the left of $z = 0$ is 0.500 square unit. 0.497 + 0.500 = 0.997.
    The area of the distribution where $z < 2.71$ is 0.997 square unit.

25. If 0.200 square unit of the area of the distribution is to the right of z, then 0.300 square unit is between z and 0.
At $A = 0.300$, $z = 0.84$.
The z-score is 0.84.

27. If 0.184 square unit of the area of the distribution is to the left of z, then 0.316 square unit is between z and 0.
At $A = 0.316$, $z = 0.90$. Because z is to the left of 0, the z-score is –0.90.

29. If 0.363 square unit of the area of the distribution is to the right of z, then 0.137 square unit is between z and 0.
At $A = 0.137$, $z = 0.35$.
The z-score is 0.35.

31. a. $z_{219} = \dfrac{219-185}{39} = 0.87$

Table 8.9 indicates that 0.308 (30.8%) of the data in a normal distribution are between $z = 0$ and $z = 0.87$.
Therefore, the percent of data to the right of $z = 0.87$ is 50% – 30.8% = 19.2%.
Thus 19.2% of the women's cholesterol level is greater than 219.

b. $z_{190} = \dfrac{190-185}{39} = 0.13$

Table 8.9 indicates that 0.052 (5.2%) of the data in a normal distribution are between $z = 0$ and $z = 0.13$.

$z_{225} = \dfrac{225-185}{39} = 1.03$

Table 8.9 indicates that 0.348 (34.8%) of the data in a normal distribution are between $z = 0$ and $z = 1.03$.
Thus the percent of women's cholesterol levels between 190 and 225 is 34.8% – 5.2% = 29.6%.

33. a. $z_{950} = \dfrac{950-1025}{87} = -0.86$

Table 8.9 indicates that 0.305 (30.5%) of the data in a normal distribution are between $z = -0.86$ and $z = 0$. Therefore, the percent of data to the right of $z = -0.86$ is 50% + 30.5% = 80.5%. Thus 80.5% of the light bulb's life span is at least 950 hours.

b. $z_{800} = \dfrac{800-1025}{87} = -2.59$

Table 8.9 indicates that 0.495 (49.5%) of the data in a normal distribution are between $z = -2.59$ and $z = 0$.

$z_{900} = \dfrac{900-1025}{87} = -1.44$

Table 8.9 indicates that 0.425 (42.5%) of the data in a normal distribution are between $z = -1.44$ and $z = 0$.
Thus the percent of the light bulb's life span between 800 and 900 hours is 49.5% – 42.5% = 7%.

35. a. $z_{14} = \dfrac{14-14.5}{0.4} = -1.25$

0.394 square unit is between $z = -1.25$ and $z = 0$. Therefore, 0.106 square unit is to the left of $z = -1.25$.
10.6% of the boxes will weigh less than 14 ounces.

b. Calculating:

$z_{13.5} = \dfrac{13.5-14.5}{0.4} = -2.5$

$z_{15.5} = \dfrac{15.5-14.5}{0.4} = 2.5$

0.494(2) = 0.988, so 98.8% of the boxes will weigh between 13.5 and 15.5 oz.

37. a. $z_{320} = \dfrac{320-350}{24} = -1.25$

0.394 square unit is between $z = -1.25$ and $z = 0$. Therefore, 0.106 square unit is to the right of $z = -1.25$.
The probability that a piece of this rope chosen at random will have a breaking point of less than 320 pounds is 0.106.

b. Calculating:

$z_{340} = \dfrac{340-350}{24} \approx -0.42$

$z_{370} = \dfrac{370-350}{24} \approx 0.83$

0.163 square unit is between $z = -0.42$ and $z = 0$. 0.297 square unit is between $z = 0$ and $z = 0.83$. 0.163 + 0.297 = 0.460.
The probability of a piece of this rope chosen at random will have a breaking point of between 340 and 370 pounds is 0.460.

39. a. $z_3 = \dfrac{3-2.5}{0.75} \approx 0.67$

0.249 square unit is between $z = 0$ and $z = 0.67$. 0.500 square unit is to the left of $z = 0$. 0.249 + 0.500 = 0.749.
The probability that the time a customer spends waiting is at most 3 minutes is 0.749.

b. $z_1 = \dfrac{1-2.5}{0.75} = -2$

0.477 square unit is between $z = -2$ and $z = 0$. Therefore, 0.023 square unit is left of $z = -2$. The probability that the time a customer spends waiting is less than 1 minute is 0.023.

41. Answers will vary.

CHAPTER 8 REVIEW EXERCISES

1. $\dfrac{12+17+14+12+8+19+21}{7} \approx 14.7 = \mu$

The mean of the population is about 14.7.
The median is the fourth ranked number, or 14.
The mode is the most common number, or 12.
The range is 21 – 8 = 13.

x	$x-\mu$	$(x-\mu)^2$
12	12 – 14.7	$(-2.7)^2 = 7.29$
17	17 – 14.7	$(2.3)^2 = 5.29$
14	14 – 14.7	$(-0.7)^2 = 0.49$
12	12 – 14.7	$(-2.7)^2 = 7.29$
8	8 – 14.7	$(-6.7)^2 = 44.89$
19	19 – 14.7	$(4.3)^2 = 18.49$
21	21 – 14.7	$(6.3)^2 = 39.69$
		123.43

$\dfrac{\text{sum of the (deviations)}^2}{n} = \dfrac{123.43}{7} \approx 17.6$

The standard deviation of the population is $\sigma = \sqrt{17.6} \approx 4.2$.
The variance of the population is $\sigma^2 = (\sqrt{17.6})^2 \approx 17.6$.

2. 14, as the mode is the most common number.

3. Answers will vary.

4. a. The median.

   b. The mode.

   c. The mean.

5. (1235 + 1644 + 1576 + 1200 + 1200 + 1182 + 1212 + 1400) / 8 ≈ 1331.125
The mean is about 1331.125 feet.
The median is the average of the fourth and fifth ranked numbers, or 1223.5 feet.
The mode is the most common number, or 1200 feet.
The range is the difference between the highest and lowest values, or 462 feet.

6. $\dfrac{90}{2.5} = 36$ The average rate was 36 mph.

7. [3(4) + 3(2.33) + 2(2.67) + 4(3) + 1(4)]/13 ≈ 3.10.
The student has a grade point average of about 3.10.

8. a. $z_{82} = \dfrac{82-72}{8} = 1.25$
Ann's z-score is 1.25

   b. Ann's percentile $= \dfrac{35}{40} \cdot 100 = 87.5$
Ann was at the 88th percentile.

9. Mean $= \dfrac{12+18+20+14+16}{5} = 16$

Sample variance $= \dfrac{\sum(x-\bar{x})^2}{n-1}$

$= \dfrac{(12-16)^2+(18-16)^2+(20-16)^2+(14-16)^2+(16-16)^2}{4}$

$= 10$

$\sqrt{10} \approx 3.16$

The sample variance is 10 minutes². The sample standard deviation is 3.16 minutes.

10. Mean

$= \dfrac{6.55+6.88+7.18+7.50+7.89+7.93+7.96+8.15+8.17+8.12}{10}$

$\approx 7.63 = \bar{x}$

The sample mean is about \$7.63.

Median is $\dfrac{7.89+7.93}{2} = \$7.91$.

x	$x-\bar{x}$	$(x-\bar{x})^2$
6.55	6.55 – 7.63 = –1.08	1.1664
6.88	6.88 – 7.63 = –0.75	0.5625
7.18	7.18 – 7.63 = –0.45	0.2025
7.50	7.50 – 7.63 = –0.13	0.0169
7.89	7.89 – 7.63 = 0.26	0.0676
7.93	7.93 – 7.63 = 0.30	0.09
7.96	7.96 – 7.63 = 0.33	0.1089
8.15	8.15 – 7.63 = 0.52	0.2704
8.17	8.17 – 7.63 = 0.54	0.2916
8.12	8.12 – 7.63 = 0.49	0.2401
	Total =	3.0169

Divide the sum by n – 1:

$\dfrac{3.0169}{10-1} \approx 0.33521$

Standard deviation for this sample is:

$\sqrt{0.33521} \approx 0.58$

11. a. The second student's mean is 5 points higher than the first student's mean.

    b. The two standard deviations are the same.

12. a. $z_{72} = \dfrac{72-81}{5.2} \approx -1.73$

    b. $z_{84} = \dfrac{84-81}{5.2} \approx 0.58$

13. Minimum: 164, maximum: 310
$Q_1 = 172$, $Q_2 = 196.5$, $Q_3 = 221$
Drawing the plot:

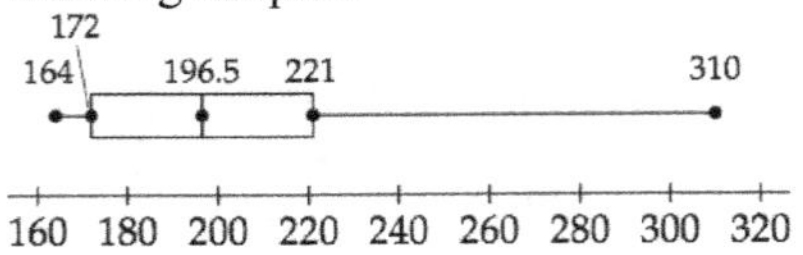

14. a. 8 students scored at least an 84 on the test.

b. 40 students took the test.

15. a. 28 + 8 + 18 + 4 + 8 + 4 + 6 + 2 = 78.
78% of the states paid an average teacher salary of at least $48,000.

b. 18 + 4 + 8 + 4 = 34. The probability that a state selected at random paid an average teacher salary of at least $56,000 but less than $72,000 is 0.34.

16. Using a calculator, the linear correlation coefficient (r) is –0.9004, which indicates a strong linear correlation.

17. a. $z_8 = \dfrac{8-6.5}{1} = 1.5$
0.433 square unit lies between $z = 1.5$ and $z = 0$.
0.500 square unit lies left of $z = 0$. 0.433 + 0.500 = 0.933.
The probability that a customer spends at most 8 minutes waiting is 0.933.

b. $z_8 = \dfrac{6-6.5}{1} = -0.5$
0.191 square unit lies between $z = -0.5$ and $z = 0$. Therefore, 0.309 square unit lies to the left of $z = -0.5$.
The probability that a customer spends less than 6 minutes waiting is 0.309.

18. a. $z_{49.5} = \dfrac{49.5-50}{0.5} = -1$
0.288 square units lie between $z = -1$ and $z =0$. Therefore, 0.341 square units lies to the left of $z = -1$, and the percent is 50% – 34.1% = 15.9%.
Thus 16% of the sacks will weigh less than 49.5 pounds.

b. 95% of the sacks will weigh between 49 and 51 pounds, as 49 and 51 are two standard deviations below and above the mean, respectively.

19. a. 0.433 square unit lies between $z = 0$ and $z =1.5$. Therefore, 0.067 square unit lies to the right of $z = 1.5$.
6.7% of the manufacturer's telephones will last at least 7.25 years.

b. Calculating:
$z_{5.8} = \dfrac{5.8-6.5}{0.5} = -1.4$
$z_{6.8} = \dfrac{6.8-6.5}{0.5} = 0.6$
0.419 square unit lies between $z = -1.4$ and $z = 0$.
0.226 square unit lies between $z = 0$ and $z = 0.6$. 0.419 + 0.226 = 0.645.
64.5% of the manufacturer's telephones will last between 5.8 and 6.8 years.

c. $z_{6.9} = \dfrac{6.9-6.5}{0.5} = 0.8$
0.288 square unit lies between $z = 0$ and $z = 0.8$.
0.500 square unit lies to the left of $z = 0$.
0.288 + 0.500 = 0.788.
78.8% of the company's phones will last less than 6.9 years.

20. (91.4 + 92.6 + 94.5 + 94.6 + 94.3 + 91.5) / 6 = 93.15 million miles

Chapter 9: Voting and Apportionment

Think About It, Section 9.1

1. It is the number of citizens represented by each representative.

2. Hamilton, Jefferson

3. Each representative represents 25,000 citizens living in that state.

4. The number of representatives apportioned to a state is the standard quota or one more than the standard quota.

5. An apportionment procedure in which an increase in the total number of representatives for a legislature would cause a district represented by the legislature to lose a seat.

Section Exercises 9.1

1. To calculate the standard divisor, divide the total population p by the number of items to apportion n.

3. The standard quota for a state is the whole number part of the quotient of the state's population divided by the standard divisor.

5. a. $\frac{1547-1215}{1215}=0.273$

   b. $\frac{1498-1195}{1195}=0.254$

   c. Salinas

7.

	Summer Hill Average	Seaside Average	Abs. Unfair
Summer Hill receives assoc.	$\frac{5289}{587+1}=8.99$	$\frac{6215}{614}=10.12$	1.13
Seaside receives assoc.	$\frac{5289}{587}=9.01$	$\frac{6215}{614+1}=10.11$	1.10

Relative unfairness if Summer Hill receives the new associate: $\frac{1.13}{8.99}=0.126$

Relative unfairness if Seaside receives the new associate: $\frac{1.10}{10.11}=0.109$

Using the apportionment principle, Seaside Mall should get the new associate.

9. a. $\frac{p}{n}=\frac{308,745,538}{435}\approx 709,760$

   There is one representative for every 709,760 citizens in the U.S.

   b. Underrepresented; the average constituency is greater than the standard divisor.

   c. Over-represented; the average constituency is less than the standard divisor.

11. a. $\frac{p}{n}=\frac{1781}{48}\approx 37.10$

   There is one new nurse for every 37.10 beds.

   b.

Hospital	Quot.	Q	SQ	beds
Sharp	$\frac{242}{37.1}$	6.52	6	7
Palomar	$\frac{356}{37.1}$	9.60	9	10
Tri-City	$\frac{308}{37.1}$	8.30	8	8
Del Raye	$\frac{190}{37.1}$	5.12	5	5
R. Verde	$\frac{275}{37.1}$	7.41	7	7
Bel Aire	$\frac{410}{37.1}$	11.05	11	11

   c. The modified standard divisor is 34.5.

Hospital	Quot.	Q	SQ	beds
Sharp	$\frac{242}{34.5}$	7.01	7	7
Palomar	$\frac{356}{34.5}$	10.32	10	10
Tri-City	$\frac{308}{34.5}$	8.93	8	8
Del Raye	$\frac{190}{34.5}$	5.51	5	5
R. Verde	$\frac{275}{34.5}$	7.97	7	7
Bel Aire	$\frac{410}{34.5}$	11.88	11	11

   d. They are identical.

13. The population paradox occurs when the population of one state is increasing faster than that of another state, yet the first state still loses a representative.

15. The Balinski-Young Impossibility Theorem states that any apportionment method will either violate the quota rule or will produce paradoxes such as the Alabama paradox.

17. a. The standard divisor is 6.57.

Resort	A	B	C	D
Guest rooms	23	256	182	301
@ 115	4	39	27	45
@ 116	3	39	28	46

Yes. Resort A will lose a TV.

b. The standard divisor is 6.51.

@ 117	4	39	28	46

No. No resort loses a TV.

c. The standard divisor is 6.46.

@ 118	3	40	28	47

Yes. Resort A will lose a TV.

19. a. The standard divisor is 61.9.

Office	Boston	Chicago
Employees	151	1210
Vice presidents	2	20

b. The standard divisor is 62.3.

Office	Boston	Chicago	SF
Employees	151	1210	135
Quota	2.424	19.422	2.167
VPs	3	19	2

Yes. Chicago lost a vice president and Boston gained one.

21. a.

	avg. constit. in 5th grade	avg. constit. in 6th grade
Add to 5th grade	$\frac{604}{19+1}=30.2$	$\frac{698}{21}\approx 33.2$
Add to 6th grade	$\frac{604}{19}\approx 31.8$	$\frac{698}{21+1}\approx 31.7$

The absolute unfairness of the apportionment if the 5th grade gets the new teacher is 33.2–30.2; the absolute unfairness of the apportionment if the sixth grade gets the new teacher is
31.8 – 31.7.
The relative unfairness of the apportionment for each grade equals the absolute unfairness of the apportionment for the grade divided by the average constituency if the grade gets the new teacher. If the fifth grade gets the new teacher, the relative unfairness of the apportionment is
(33.2–30.2)/30.2 ≈ 0.0993.
If the sixth grade gets the new teacher, the relative unfairness of the apportionment is
(31.8 – 31.7)/31.7 ≈ 0.0032.
Because the smaller relative unfairness results from adding the teacher to the sixth grade, sixth grade should get the teacher.

b. The Huntington-Hill number for the fifth grade is

$$\frac{(S_5)^2}{t(t+1)}=\frac{(604)^2}{(19)(20)}\approx 960.$$

The Huntington-Hill number for the sixth grade is

$$\frac{(698)^2}{(21)(22)}\approx 1054.6.$$

The grade with the greater Huntington-Hill number, which is grade 6, receives the new teacher. This is the same result as using the relative unfairness of the apportionment.

23.

School	Computers	Students	H-H number
Rose	26	625	556
Lincoln	22	532	559
Midway	26	620	548
Valley	31	754	573

Because Valley School has the greatest Huntington-Hill number, it should be assigned the new computer.

25. New York's population is much larger than Louisiana's, so New York has many more representatives, and a percentage change in its population adds or subtracts more representatives than for smaller states.

27.

a. The modified standard divisor is 88.

Div.	Quot.	Q	SQ	PCs
Lib. Arts	3455/88	39.26	39	39
Bus.	5780/88	65.68	66	66
Hum.	1896/88	21.55	22	22
Sci.	4678/88	53.16	53	53

This is the same result as the Hamilton method.

b. Using the Jefferson method, the humanities division gets one less computer and the sciences division gets one more computer, compared with using the Webster method.

29. The Jefferson and Webster methods

31. The Huntington-Hill method

THINK ABOUT IT, SECTION 9.2

1. An issue that passes by a majority of votes cast receives more than 50% of the votes cast. An issue that wins with a plurality of votes cast receives more votes than any other issue but not necessarily more than 50%.

2. There were five people who ranked the colors Red–3, Green–2, and Blue–1.

3. There is no voting method involving three or more choices that satisfies all four fairness criteria.

4. A candidate who wins all possible head-to-head matchups should win an election when all candidates appear on the ballot.

SECTION EXERCISES 9.2

1. There is a tie between Corn Flakes and Raisin Bran- 6 first-place votes each.

3. There is a tie between Italian and Mexican.

5. Corn Flakes: $(6 \cdot 3) + (5 \cdot 2) + 5 = 33$
Raisin Bran: $(6 \cdot 3) + (5 \cdot 2) + 5 = 33$
Mini Wheat: $(4 \cdot 3) + (6 \cdot 2) + 6 = 30$
Tie between Corn Flakes and Raisin Bran.

7. Italian: $(6 \cdot 3) + (1 \cdot 2) + 8 = 28$
Mexican: $(6 \cdot 3) + (5 \cdot 2) + 4 = 32$
Japanese: $(3 \cdot 3) + (9 \cdot 2) + 3 = 30$
The committee prefers Mexican food.

9. AT&T : $0 + (42 \cdot 3) + (81 \cdot 2) + 77 = 365$
Sprint : $(56 \cdot 4) + 0 + (39 \cdot 2) + 105 = 407$
Verizon : $(81 \cdot 4) + (119 \cdot 3) + 0 + 0 = 681$
T-Mobile : $(63 \cdot 4) + (39 \cdot 3) + (80 \cdot 2) + 18 = 547$
Verizon is the favorite provider of this group.

11. WSTR has no first-place votes, and is eliminated. The preference schedule is now:

WNNX	2	1	1	2	3
WKLS	1	3	3	1	2
WWVV	3	2	2	3	1
votes	57	72	38	61	15

WWVV has the least first-place votes (15) and is eliminated. The preference schedule is now:

WNNX	2	1	1	2	2
WKLS	1	2	2	1	1
votes	57	72	38	61	15

WNNX has 110 first-place votes, and WKLS has 133 first-place votes. The students' favorite radio station is WKLS.

13. AT&T has no first-place votes, and is eliminated. The preference schedule is now:

Nextel	1	1	3	3	3
Verizon	2	2	1	2	1
T-Mobile	3	3	2	1	2
votes	18	38	42	63	39

Nextel has the fewest first place votes (56) and is eliminated. The preference schedule is now:

Verizon	1	1	1	2	1
T-Mobile	2	2	2	1	2
votes	18	38	42	63	39

Verizon has more first-place votes (137 to 63) and is the preferred provider for this group of customers.

15. a. The amusement park, which has 28 first-place votes.

b. The dinner cruise received no first-place votes and is eliminated first. The preference schedule is now:

picnic	1	2	1	2	4
ski	3	1	2	3	3
park	2	4	4	1	2
horses	4	3	3	4	1
votes	10	18	6	28	16

Picnicking and horse-back riding are tied for fewest first-place votes, with 16. Because horse-riding has no second-place votes while picnicking does, horse-back riding is eliminated.
The preference schedule is now:

picnic	1	2	1	2	3
ski	3	1	2	3	2
park	2	3	3	1	1
votes	10	18	6	28	16

Picnicking still has the fewest first-place votes, 16, and is eliminated. (We could have eliminated it in the previous step by noticing that eliminating horse-back riding did not give picnicking any first-place votes.)

ski	2	1	1	2	2
park	1	2	2	1	1
votes	10	18	6	28	16

Water-skiing has 24 first-place votes, and the amusement park has 54 first-place votes. The employees prefer the amusement park.

c. Picnic:
$(16 \cdot 5) + (18 \cdot 4) + (28 \cdot 3) + (16 \cdot 2) + 0$
$= 268$
Water-skiing:
$(18 \cdot 5) + (6 \cdot 4) + (26 \cdot 3) + (28 \cdot 2) + 0$
$= 248$
Amusement Park:
$(28 \cdot 5) + (26 \cdot 4) + 0 + 0 + 24 = 268$
Riding Horses:
$(16 \cdot 5) + 0 + (6 \cdot 3) + (18 \cdot 2) + 38 = 172$
Dinner Cruise:
$0 + (28 \cdot 4) + (18 \cdot 3) + (16 \cdot 2) + 16 = 214$
There is a tie between a picnic in the park and the amusement park.

17.

versus	Grand Canyon	Yosemite	Bryce Canyon	Yellowstone
Grand Canyon	--	Grand Canyon	Grand Canyon	Grand Canyon
Yosemite	--	--	Bryce Canyon	Yosemite
Bryce Canyon	--	--	--	Bryce Canyon
Yellowstone	--	--	--	--

Total votes = 43

The Grand Canyon is preferred over Yosemite on 36 ballots, against 7 for Yosemite. The Grand Canyon is preferred over Bryce Canyon on 23 ballots, against 20 for Bryce. The Grand Canyon is preferred on 28 ballots, against 15 for Yellowstone. Yosemite is preferred over Bryce Canyon on 10 ballots, against 33 for Bryce. Yosemite is preferred over Yellowstone on 22 ballots, against 21 for Yellowstone. Bryce is preferred over Yellowstone on 25 ballots, against 18 for Yellowstone.

The Grand Canyon wins three matchups, Bryce Canyon wins two, and Yosemite wins one. The family prefers the Grand Canyon for the reunion.

19.

versus	Gibson	Yung	Allenbaugh	Meckley	DeWitte
Gibson	--	Gibson	Allenbaugh	Gibson	Gibson
Yung	--	--	Allenbaugh	Meckley	DeWitte
Allenbaugh	--	--	--	Allenbaugh	DeWitte
Meckley	--	--	--	--	DeWitte
DeWittte	--	--	--	--	--

Total votes = 52

Gibson has 36 ballots versus 16 for Yung; Gibson has 22 ballots versus 30 for Allenbaugh; Gibson has 32 ballots versus 20 for Meckley; Gibson has 36 ballots versus 16 for DeWitte; Yung has 20 ballots versus 32 for Allenbaugh; Yung has 16 ballots versus 36 for Meckley; Yung has 13 ballots versus 39 for DeWitte; Allenbaugh has 48 ballots versus 4 for Meckley; Allenbaugh has 23 ballots versus 29 for DeWitte; Meckley has 13 ballots versus 39 for DeWitte.

Gibson, Allenbaugh, and DeWitte are in a three-way tie for first.

21.

versus	WNNX	WKLS	WWVV	WSTR
WNNX	--	WKLS	WNNX	WNNX
WKLS	--	--	WWVV	WKLS
WWVV	--	--	--	WWVV
WSTR	--	--	--	--

Total votes = 243

WNNX has 110 votes versus 133 for WKLS; WNNX has 228 votes versus 15 for WWVV; WNNX has 171 votes versus 72 for WSTR; WKLS has 118 votes versus 125 for WWVV; WKLS has 205 votes versus 38 for WSTR; WWVV has 186 votes versus 57 for WSTR.

WNNX, WKLS, and WWVV are in a three-way tie for first.

23.

versus	Italian	Mexican	Japanese
Italian	--	Mexican	Japanese
Mexican	--	--	Mexican
Japanese	--	--	--

No. Using plurality voting, Italian and Mexican tied, although Mexican won both head-to-head contests.

25.

versus	Picnic	Ski	Amusement	Horse Riding	Cruise
Picnic	--	picnic	amusement	picnic	picnic
Ski	--	--	amusement	ski	ski
Amusement	--	--	--	amusement	amusement
Horse Riding	--	--	--	--	cruise
Cruise	--	--	--	--	--

No. Using the Borda count method, a picnic in the park and the amusement park tied, although the amusement park won all of its head-to-head contests.

27. No. Using plurality with elimination, WKLS won the vote, although it fell a few votes short of a majority.

29. a. Film C has the least first-place votes (5) and is eliminated. The preference schedule is now:

Film A	1	2	1	1
Film B	2	1	2	2
votes	4	6	5	2

Film A has 11 first-place votes, and Film B has 6, so Film A wins the competition.

b. On the revote, film A has the fewest first-place votes (4) and is eliminated. The preference schedule is now:

Film B	1	1	2	2
Film C	2	2	1	1
votes	4	6	5	2

Now, Film B has 10 first-place votes and Film C has 7 first-place votes. Film B wins the competition.

c. No. The montonicity criterion requires that, if a candidate wins an election that candidate should also win the election if the only change is that supporters of another candidate change their votes to support the winning candidate. While Film A won the first election, no voters switched their votes to Film A for the revote.

THINK ABOUT IT, SECTION 9.3

1. A weighted voting system is a voting system in which some voters have more influence on the outcome of an election.

2. A quota in a weighted voting system is the number of votes required to pass a measure or elect a person to office.

3. If one of the permanent members votes against a resolution, there are only 38 votes remaining even if every other country voted for the resolution. This number is less than the quota, 39.

4. No. If one of the permanent members votes against a resolution, the four remaining members vote for it, and all the elected members vote for it, the measure would pass.

SECTION EXERCISES 9.3

1. a. 6
   b. 4
   c. 3
   d. 6
   e. No
   f. A and C
   g. $2^4 - 1 = 15$
   h. 6

3.

Winning coalition	Critical voters
{A, B}	A, B
{A, C}	A, C
{A, B, C}	A

$BPI(\text{A}) = \frac{3}{5} = 0.6$

$BPI(\text{B}) = BPI(\text{C}) = \frac{1}{5} = 0.2$

5.

Winning coalition	Critical voters
{A, B}	A, B
{A, B, C}	A, B
{A, B, D}	A, B
{A, C, D}	A, C, D
{A, B, C, D}	A

$BPI(\text{A}) = \frac{5}{10} = 0.5$

$BPI(\text{B}) = \frac{3}{10} = 0.3$

$BPI(\text{C}) = \frac{1}{10} = 0.1 = BPI(\text{D})$

7.

Winning coalitions	Critical voters
{A, B}	A, B
{A, B, C}	A, B
{A, B, D}	A, B
{A, B, E}	A, B
{A, C, D}	A, C, D
{A, C, E}	A, C, E
{B, C, D}	B, C, D
{A, B, C, D}	none
{A, B, C, E}	A
{A, B, D, E}	A, B
{A, C, D, E}	A, C
{B, C, D, E}	B, C, D
{A, B, C, D, E}	none

$BPI(\text{A}) = \frac{9}{25} = 0.36;\ BPI(\text{B}) = \frac{7}{25} = 0.28$

$BPI(\text{C}) = \frac{5}{25} = 0.20;\ BPI(\text{D}) = \frac{3}{25} = 0.12$

$BPI(\text{E}) = \frac{1}{25} = 0.04$

9. No coalition without voter A can reach a quota. Voter A has a Banzhaf power index of 1; all the other voters have a BPI of 0.

11.

Winning Coalition	Critical Voters
{A, B}	A, B
{A, C}	A, C
{A, B, C}	A
{A, B, D}	A, B
{A, B, E}	A, B
{A, C, D}	A, C
{A, C, E}	A, C
{A, D, E}	A, D, E
{B, C, D}	B, C, D
{A, B, C, D}	none
{A, B, C, E}	A
{A, B, D, E}	A
{A, C, D, E}	A
{B, C, D, E}	B, D, C
{A, B, C, D, E}	none

$$BPI(\text{A}) = \frac{11}{25} = 0.44$$

$$BPI(\text{B}) = BPI(\text{C}) = \frac{5}{25} = 0.20$$

$$BPI(\text{D}) = \frac{3}{25} = 0.12$$

$$BPI(\text{E}) = \frac{1}{25} = 0.04$$

13. a. In Exercise 9, voter A never needs any of the other voters in any winning coalition, and is thus the dictator.

    b. Veto power means that one or more voters are critical in every winning coalition. In exercises 3 and 5, voter A is critical in each winning coalition. In Exercise 6, there is only one coalition, so each voter has veto power. In Exercise 9, voter A is a dictator and thus has veto power. In exercise 12, voter A is critical in all winning coalitions.

    c. None

    d. In Exercise 8, each voter has one vote.

15. The winning coalitions are {D, T}, {D,P}, and {T, P}, with each voter critical to each. The director, teacher, and principal each have a Banzhaf power index of 0.33.

17. a. {12: 1, 1, 1, 1, 1, 1, 1, 1, 1, 1, 1, 1,}

    b. Yes.

    c. Yes, every juror has a veto, as the verdict must be unanimous.

    d. Because this is a veto power system, divide the voter power of 1 by the quota of 12.

19. A is the dictator, and all of B, C, D, and E are dummies.

21. None.

23. a. The winning coalitions are {A, B}, {A, C}, {A, B, C}. A is critical to all three (and has veto power), and B and C are each critical in their own coalition with A. The BPI for A is 0.6, and for B and C the BPI is 0.2

    b. Answers will vary.

25. a. There are four winning coalitions: {A, B}, {A, C}, {B, C}, and {A, B, C}. The critical voters are, respectively, A and B, A and C, B and C, and none of them. Each voter has a BPI of 0.33.

    b. This system has the same effect as a one-person, one-vote system.

CHAPTER 9 REVIEW EXERCISES

1. a. The standard divisor is $\frac{9606}{50} = 192.18$.

Div.	Enrol.	Quota	Proj.
Health	1280	6.66	7
Bus.	3425	17.82	18
Eng.	1968	10.24	10
Sci.	2936	15.28	15

Health division gets 7 projectors, the Business division gets 18, the Engineering division gets 10, and the Science division gets 15.

   b. The modified standard divisor is 183.4.

Div.	Enrol.	Quota	Proj.
Health	1280	6.98	6
Bus.	3425	18.68	18
Eng.	1968	10.73	10
Sci.	2936	16.01	16

   c. A modified standard divisor of 191 yields the following apportionment.

Div.	Enrol.	Quota	Proj.
Health	1280	6.70	7
Bus.	3425	17.93	18
Eng.	1968	10.30	10
Sci.	2936	15.37	15

2. a. The standard divisor is $\frac{110}{35} = 3.143$.

Airport	Agts.	Quota	Empl.
Newark	28	8.91	9
Cleve.	19	6.05	6
Chi.	34	10.82	11
Phila.	13	4.14	4
Det.	16	5.09	5

Newark gets 9 security employees, Cleveland 6, Chicago 11, Philadelphia 4, and Detroit 5.

b. The modified standard divisor is 3.05.

Airport	Agts.	Quota	Empl.
Newark	28	9.18	9
Cleve.	19	6.23	6
Chicago	34	11.15	11
Phila.	13	4.26	4
Detroit	16	5.25	5

c. The same modified standard divisor of 3.05 yields the Webster apportionment.

Airport	Agts.	Quota	Empl.
Newark	28	9.18	9
Cleve.	19	6.23	6
Chi.	34	11.15	11
Phila.	13	4.26	4
Det.	16	5.25	5

3. a. Relative unfairness if High Desert receives the new controller: $\frac{326-297}{297} \approx 0.098$

b. Relative unfairness if Eastlake receives the new controller: $\frac{302-253}{253} \approx 0.194$.

c. Using the apportionment principle, High Desert Airport should get the new controller.

4.

	Morena Valley Average	West Keyes Average	Abs. Unfair
Morena Valley receives professor	$\frac{1437}{38+1} = 36.85$	$\frac{1504}{46} = 32.70$	4.15
West Keyes receives professor	$\frac{1437}{38} = 37.82$	$\frac{1504}{46+1} = 32.00$	5.82

Relative unfairness if Morena Valley receives the new professor: $\frac{4.15}{36.85} \approx 0.113$

Relative unfairness if Eastlake receives the new controller: $\frac{5.82}{32.00} \approx 0.182$

Morena Valley should receive the new professor.

5. a. The standard divisor is 13.79. Office A has a quota is 1.38, B has a quota of 14.14, C has a quota of 22.34, and D has a quota of 29.15. Office A gets the next printer, and no office loses a printer. The Alabama paradox does not occur.

b. The standard divisor is 13.59, and the quotas are as follows: A:1.40; B: 14.35; C: 22.66; and D: 29.58. Offices C and D get the new printers, and Office A falls back to 1 printer. This is an example of the Alabama paradox.

6. a. The standard divisor is 20.96.

Center	Quota	Autos
A	1.48	2
B	5.15	5
C	3.34	3
D	15.70	16
E	2.34	2

Regional center D gets the additional automobile. There is no Alabama paradox, as no center loses an automobile.

b. The standard divisor is 20.24.

Center	Quota	Autos
A	1.53	2
B	5.34	5
C	3.46	4
D	16.25	16
E	2.42	2

Regional center C gets the additional automobile. There is no Alabama paradox, as no center loses an automobile.

7. a. The standard divisor is 151.4. Los Angeles gets 9 servers, and Newark gets 2.

b. The standard divisor is now 148. LA gets 10, and Newark and KC get 1 each. Yes, this is the new states' paradox, since Newark loses a server without having lost people.

8. a. The standard divisor is 1408. The apportionment is A: 10, B: 3, C: 21.

b. The standard divisor is 1428. The apportionment is A: 11, B: 2, C: 21. Yes, this is the population paradox. Although the population of region B has grown at a higher rate than the population region A, region B still loses an inspector in reapportionment.

9. Yes.

10. No, it cannot occur.

11. a. The Huntington-Hill number is given by $\frac{(P_A)^2}{a(a+1)}$, where P_A is the population, and a is the current apportionment.

Bldg	A	B	C
Guard	25	43	18
Empl.	414	705	293
H-H	263.7	262.7	251.0

Building A has the greatest Huntington-Hill number, and is assigned the next guard.

b.

Bldg	**A**	**B**	**C**
Guard	26	43	18
Empl.	414	705	293
H-H	244.2	262.7	251.0

Building B now has the greatest Huntington-Hill number, and is assigned the 88th guard.

12. a. Ortega, with 8 votes.

b. No, he has 8 of 23 votes.

c. Kelly: $(2 \cdot 8)+(3 \cdot 5)+(3 \cdot 4)+(4 \cdot 6)$
$= 67$
Ortega: $(4 \cdot 8)+(2 \cdot 5)+(1 \cdot 4)+(2 \cdot 6)$
$= 58$
Nisbet: $(3 \cdot 8)+(1 \cdot 5)+(4 \cdot 4)+(3 \cdot 6)$
$= 63$
Toyama: $(1 \cdot 8)+(4 \cdot 5)+(2 \cdot 4)+(1 \cdot 6)$
$= 42$
Crystal Kelly wins the essay contest.

13. a. Vail 23, Aspen 22, Powderhorn 18 The club chooses Vail.

b. Aspen:
$(5 \cdot 14)+(5 \cdot 8)+(3 \cdot 11)+(4 \cdot 18)+(3 \cdot 12)$
$= 251$
Copper Mtn.:
$(1 \cdot 14)+(2 \cdot 8)+(4 \cdot 11)+(2 \cdot 18)+(2 \cdot 12)$
$= 134$
Powderhorn:
$(3 \cdot 14)+(4 \cdot 8)+(1 \cdot 11)+(5 \cdot 18)+(1 \cdot 12)$
$= 187$
Telluride:
$(2 \cdot 14)+(1 \cdot 8)+(2 \cdot 11)+(1 \cdot 18)+(4 \cdot 12)$
$= 124$
Vail:
$(4 \cdot 14)+(3 \cdot 8)+(5 \cdot 11)+(3 \cdot 18)+(5 \cdot 12)$
$= 249$
Using the Borda Count method, the club chooses Aspen.

14. Schneider has no first-place votes, and is eliminated. The preference schedule is now:

Reynolds	2	2	1	3
Hernandez	1	3	3	2
Kim	3	1	2	1
votes	132	214	93	119

Reynolds now has the fewest first-place votes, and is eliminated. The preference schedule is now:

Hernandez	1	2	2	2
Kim	2	1	1	1
votes	132	214	93	119

Kim wins the election, 426 to 132.

15. Twix has no first-place votes, and is eliminated. The preference schedule is now:

Crunch	1	3	3	2	2
Snickers	2	1	2	3	1
Milky Way	3	2	1	1	3
votes	15	38	27	16	22

Nestle Crunch has the fewest first-place votes, and is eliminated. The preference schedule is now:

Snickers	1	1	2	2	1
Milky Way	2	2	1	1	2
votes	15	38	27	16	22

Snickers is the favorite candy bar, 75 to 43.

16.

versus	**GR**	**LH**	**AK**	**JS**
GR	--	GR	AK	GR
LH	--	--	AK	JS
AK	--	--	--	AK
JS	--	--	--	--

Kim wins 3 matches to 2 for Reynolds and 1 for Schneider.

17.

versus	**Crunch**	**Snick.**	**Milky**	**Twix**
Crunch	--	Snick.	Milky	Twix
Snick.	--	--	Snick.	Snick.
Milky	--	--	--	Milky
Twix	--	--	--	--

Snickers is the favorite candy bar, with 3 matches compared to 2 for Milky Way and 1 for Twix.

18. a. Cynthia:
$(3 \cdot 11)+(2 \cdot 97)+(1 \cdot 112)=339$
Hannah:
$(3 \cdot 112)+(1 \cdot 97)+(1 \cdot 11)=444$
Shannon:
$(3 \cdot 97)+(2 \cdot 112)+(2 \cdot 11)=537$
Shannon is the homecoming queen.

b.

versus	Cynthia	Hannah	Shannon
Cynthia	--	Hannah	Shannon
Hannah	--	--	Hannah
Shannon	--	--	--

Hannah wins all head-to-head contests.

c. The candidate who wins all head-to-head contests should win the election of all candidates, but while Hannah wins all head-to-head contests, Shannon wins the overall election.

d. Hannah, with 112 votes, wins both the plurality and the majority of votes.

e. Hannah won the majority of first-place votes, but Shannon won the overall election.

19. a. The resulting preference schedule:

Hannah	1	2	2
Shannon	2	1	1
votes	112	97	11

Hannah wins the election, 112 to 108.

b. Cynthia L., a losing candidate, withdrew from the race and caused a change in the overall winner of the election.

20. The montonicity criterion has been violated. If a candidate wins an election, then, if the only change in voters' preferences is that supporters of another candidate change their votes to support the winner, then the result must remain the same. Margaret won the scholarship, but, when voters changed their preference from Terry to Margaret, Jean won the scholarship.

21. a. 18

b. 18

c. Yes

d. A and C

e. 15

f. 6

22. a. 35

b. 35

c. Yes

d. A

e. 31

f. 10

23. The winning coalitions are {A, B}, with A and B critical members; {A, C}, with A and C critical members; and {A, B, C}, with A as the critical member. $BPI(\text{A}) = \frac{3}{5} = 0.6$

$BPI(\text{B}) = BPI(\text{C}) = \frac{1}{5} = 0.2$

24. In a one-person, one-vote system, each voter has the same power as any other. As there are five voters in this system, each one has a BPI of $\frac{1}{5} = 0.2$.

25.

Winning coalitions	Critical voters
{A, B}	A, B
{A, C}	A, C
{A, B, C}	A
{A, B, D}	A, B
{A, C, D}	A, C
{B, C, D}	B, C, D
{A, B, C, D}	none

$BPI(\text{A}) = \frac{5}{12} \approx 0.42$

$BPI(\text{B}) = BPI(\text{C}) = \frac{3}{12} = 0.25$

$BPI(\text{D}) = \frac{1}{12} \approx 0.08$

26.

Winning Coalition	Critical Voters
{A, B}	A, B
{A, C}	A, C
{A, B, C}	A
{A, B, D}	A, B
{A, B,E}	A, B
{A, C, D}	A, C
{A, C, E,}	A, C
{A, D, E}	A, D, E
{A, B, C, D}	A
{A, B, C, E}	A
{A, B, D, E}	A
{A, C, D, E}	A
{A, B, C, D, E}	A

$BPI(\text{A}) \approx 0.62$;
$BPI(\text{B}) = BPI(\text{C}) \approx 0.14$;
$BPI(\text{D}) = BPI(\text{E}) \approx 0.05$

27. A is the dictator, and B, C, D, and E are dummies.

28. There is no dictator, and D is a dummy.

29.

Winning Coalition	Critical Voters
{A, B}	A, B
{A, C}	A, C
{A, D}	A, D
{A, B, C}	A
{A, B, D}	A
{A, C, D}	A
{A, B, C, D}	A
{A, E}	A, E
{B, C, D, E}	B, C, D, E

$BPI(\text{A}) = 0.5$;
$BPI(\text{B}) = BPI(\text{C}) = BPI(\text{D}) = BPI(\text{E}) = 0.125$

Chapter 10: Circuits and Networks

Think About It, Section 10.1

1. No. Not every possible edge is drawn. We can add two more edges:

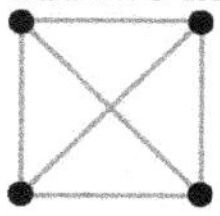

2. Answers may vary. Here is a connected graph.

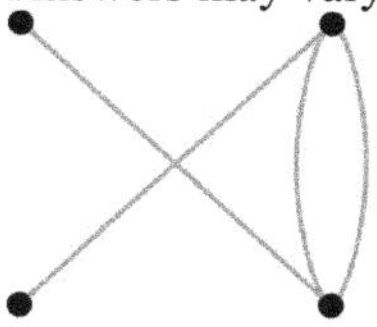

Here is a graph that is not connected.

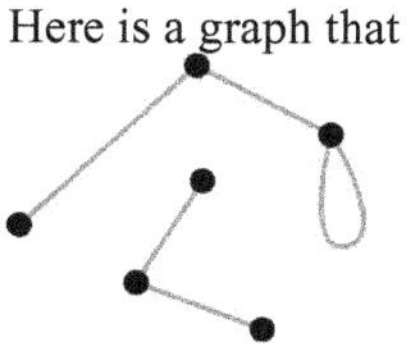

3. The number of edges that meet at a vertex is called the degree of a vertex.
4. If a path ends at the same vertex at which it started, it is considered a circuit.
5. An Euler circuit is one that travels along each edge exactly once.

Section Exercises 10.1

1. The graph is shown below:

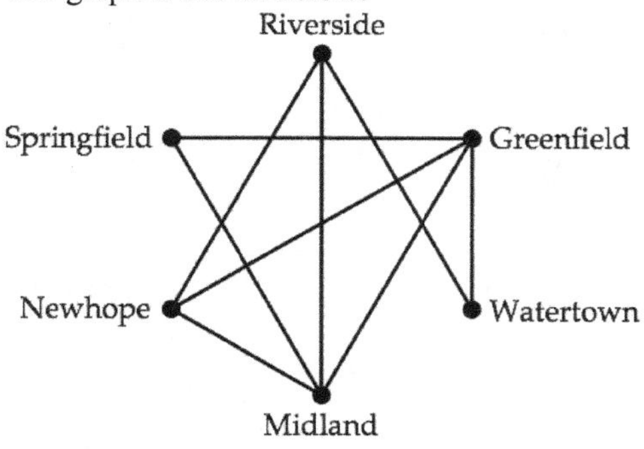

3. a. No. There is no line connecting John and Stacy.
   b. 3
   c. Ada
   d. A loop would correspond to a friend talking to himself or herself.

5. a. 6
   b. 7
   c. 6
   d. Yes
   e. No

7. a. 6
   b. 4
   c. 4
   d. Yes
   e. Yes

9. Equivalent

11. Not equivalent

13. The graph on the right has a vertex of degree 4 and the graph on the left does not.

15. a. Yes. D-A-E-B-D-C-E-D is an Euler circuit.

17. a. Not Eulerian. There are vertices of odd degree.
    b. Yes. A-E-A-D-E-D-C-E-C-B-E-B is an Euler walk.

19. a. Not Eulerian. There are vertices of odd degree.
    b. This graph does not have an Euler walk. More than two vertices are of odd degree.

21. a. Not Eulerian. There are vertices of odd degree.
    b. Yes. E-A-D-E-G-D-C-G-F-C-B-F-A-B-E-F is an Euler walk.

23. a. As a graph:

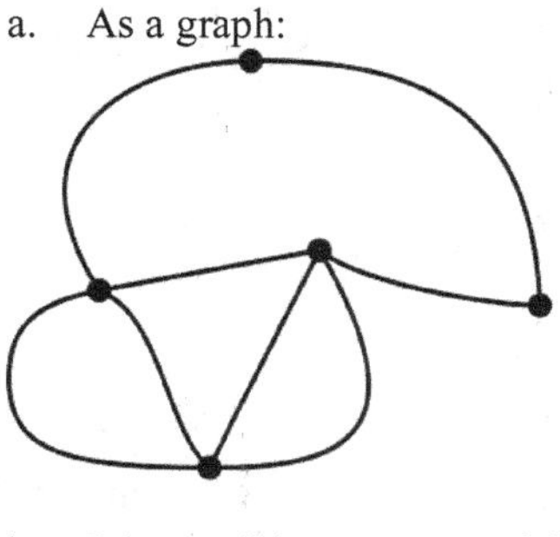

    b. It is possible to cross each bridge once and return to the starting point. Every vertex of the graph has an even degree.

25. Yes. There are exactly two vertices of odd degree, so an Euler walk is possible.

27. Yes, the hamster can travel through every tube without going through the same tube twice. Draw a graph of the Habitrail, with the vertices representing the cages and the edges representing the tubes and find an Euler circuit or an Euler walk. There is no Euler circuit; therefore the hamster cannot return to its starting point.

29. The graph is shown below:
Yes. You can always return to the starting point.

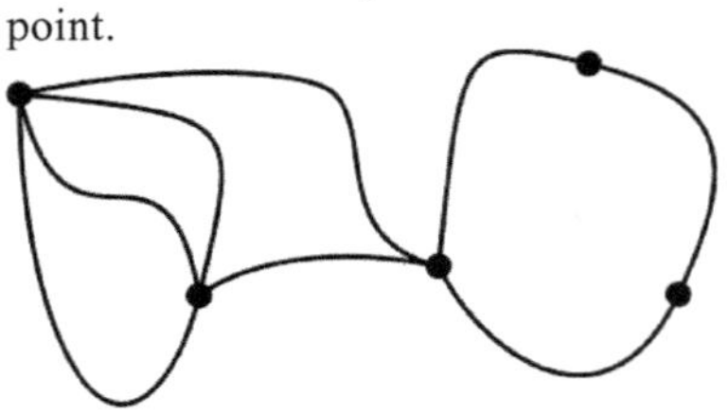

THINK ABOUT IT, SECTION 10.2

1. A Hamiltonian path passes through each vertex (note not each edge) exactly once. If it ends at the initial vertex, then it is a Hamiltonian circuit.

2. An Euler circuit is a path that crosses every edge exactly once without repeating and ends at the initial vertex. A Hamiltonian circuit passes through each vertex (note not each edge) exactly once and ends at the initial vertex.

3. A weighted graph is a graph in which each edge is associated with a value, called a weight.

4. Answers may vary. One possibility is designing the most efficient route to route packages for a package delivery company.

5. A complete graph is one in which every possible edge is drawn between vertices without any multiple edges.

SECTION EXERCISES 10.2

1. The graph is connected and has at least three vertices and no multiple edges, so the theorem applies. Every vertex has a degree of 4 or more, so the graph is Hamiltonian. A Hamiltonian circuit is A-B-C-D-E-G-F-A.

3. The graph is connected and has at least three vertices and no multiple edges, so the theorem applies. Every vertex has a degree of 4 or more, so the graph is Hamiltonian. A Hamiltonian circuit is A-B-E-C-H-D-F-G-A.

5. One such route is Springfield-Greenfield-Watertown-Riverside-Newhope-Midland-Springfield.

7. Two Hamiltonian circuits are A-B-E-D-C-A, total weight 31, and A-D-E-B-C-A, total weight 32.

9. Two Hamiltonian circuits are A-D-C-E-B-F-A, total weight 114, and A-C-D-E-B-F-A, total weight 158.

11. A-D-B-C-F-E-A

13. A-C-E-B-D-A

15. A-D-B-F-E-C-A

17. A-C-E-B-D-A

19. As a graph:

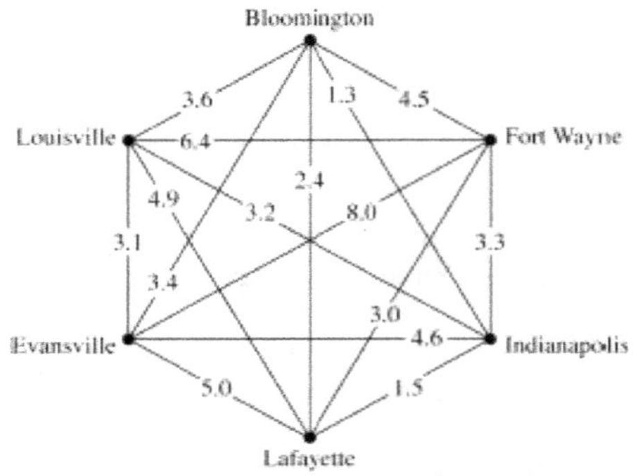

Louisville-Evansville-Bloomington-Indianapolis-Lafayette-Fort Wayne-Louisville

21. Louisville-Evansville-Bloomington-Indianapolis-Layfayette-Fort Wayne-Louisville

23. As a graph:

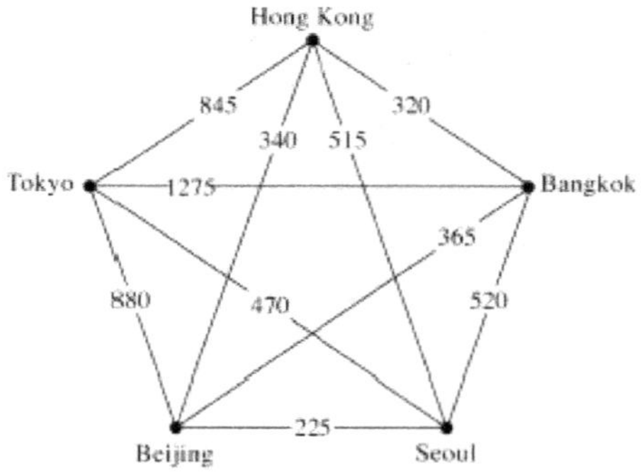

Tokyo-Seoul-Beijing-Hong Kong-Bangkok-Tokyo

25. Tokyo-Seoul-Beijing-Hong Kong-Bangkok-Tokyo

27. Represent the information as a graph:

The Greedy Algorithm generates the route home-pharmacy-pet store-farmer's market-shopping mall-home. The Edge-Picking Algorithm generates the route home-pharmacy-pet store-shopping mall-farmer's market-home.

29. Represent the information as a graph:

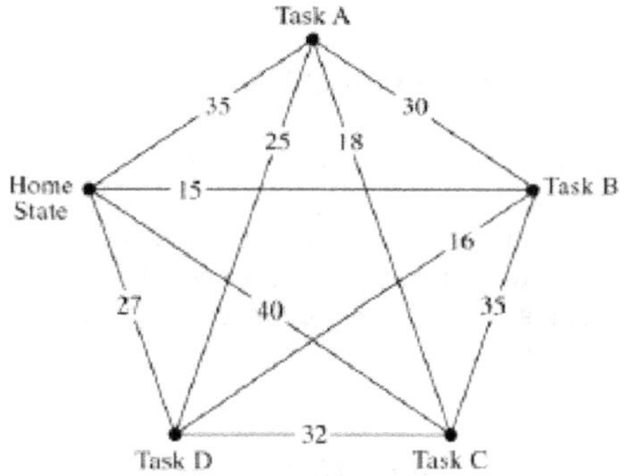

The Greedy Algorithm generates the sequence home state-task B-task D-task A-task C-home state. The Edge-Picking Algorithm gives the same sequence.

THINK ABOUT IT, SECTION 10.3

1. A planar graph is a graph that can be drawn so that no edges intersect each other (except at vertices).

2. The drawing is not planar because two edges cross. The graph is planar because we can make an equivalent planar drawing of it as shown at the right.

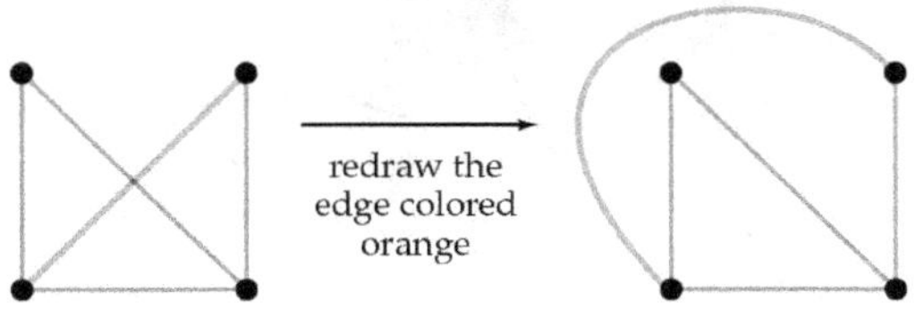

3. In a connected planar graph drawn with no intersecting edges, let v be the number of vertices, e the number of edges, and f the number of faces. Then $v + f = e + 2$.

SECTION EXERCISES 10.3

1. As a planar graph:

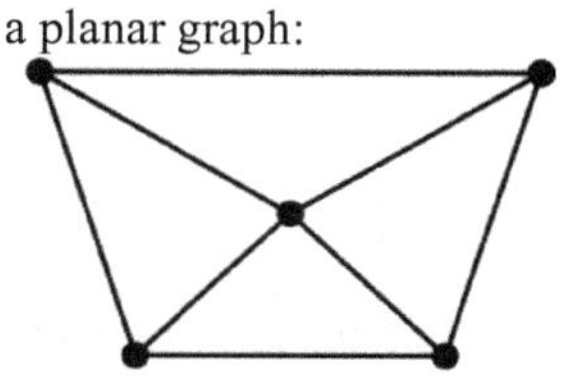

3. As a planar graph:

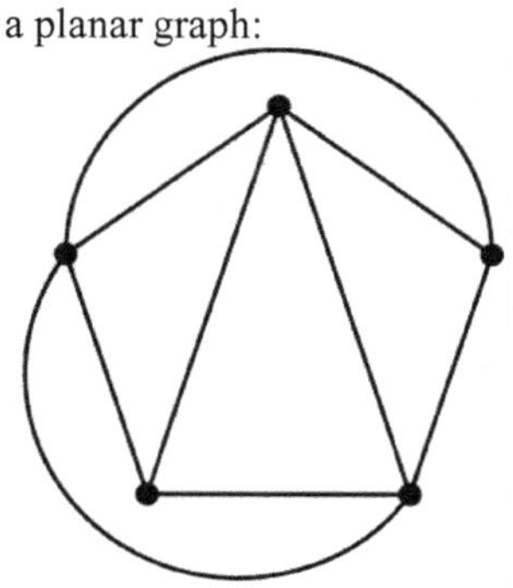

5. As a planar graph:

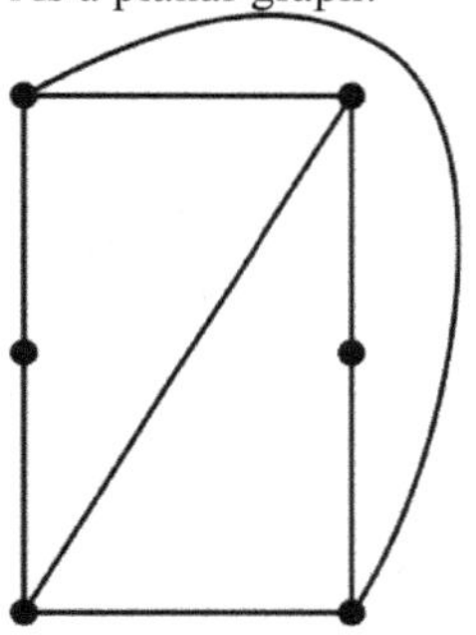

7. As a planar graph:

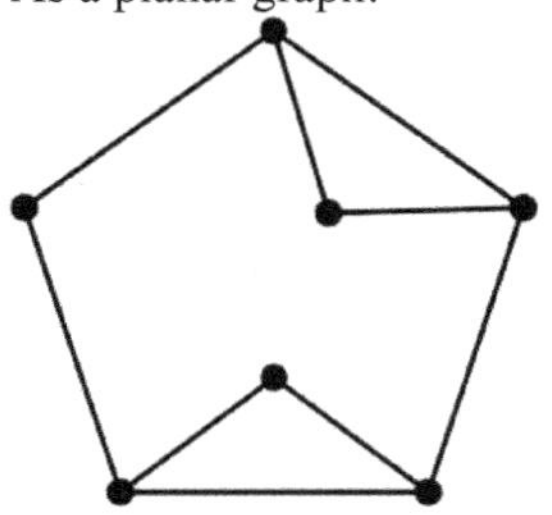

9. Finding a subgraph:

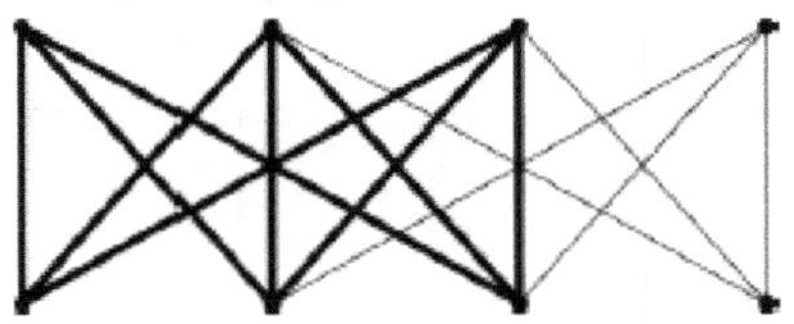

The highlighted subgraph is the Utilities Graph.

11. The graph contains K_5.

13. Starting with the original graph:

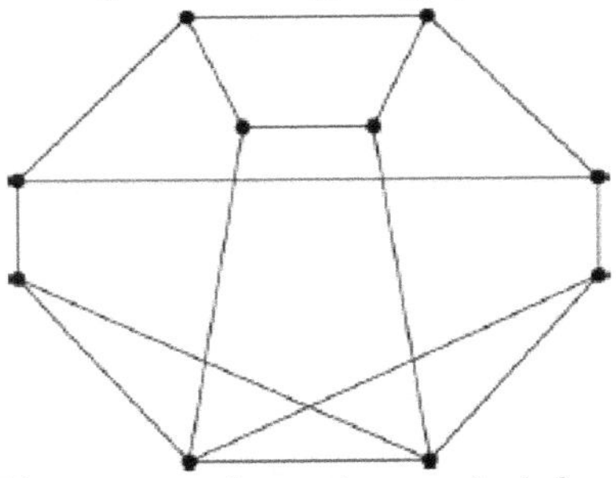

First contract the vertices on the left and right sides.

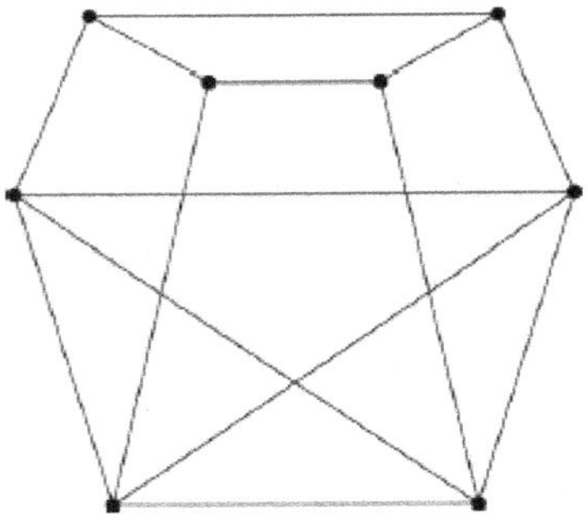

Then contract each diagonal pair of vertices at the top of the graph.

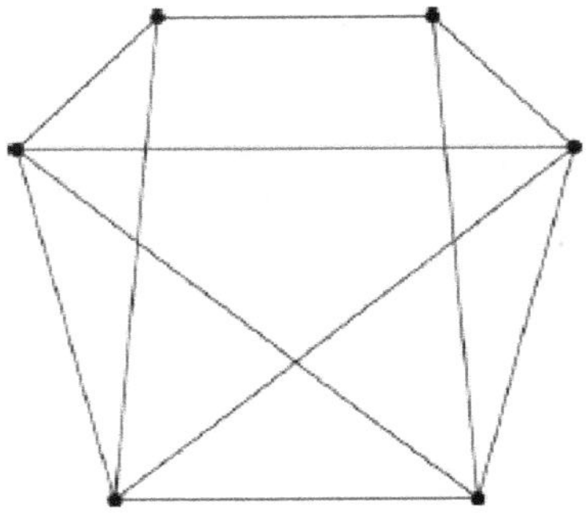

Finally contract the two vertices at the top of the graph.

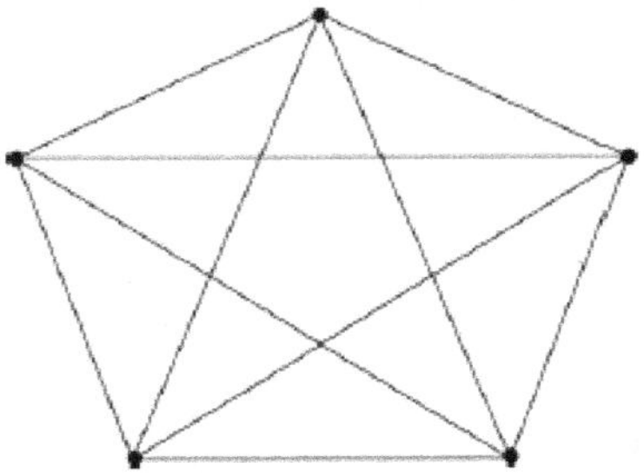

15. The graph contracts to the Utilities Graph.

17. 5 faces, 5 vertices, 8 edges

$$v + f = e + 2$$
$$5 + 5 = 8 + 2$$
$$10 = 10$$

19. 2 faces, 8 vertices, 8 edges

$$v + f = e + 2$$
$$8 + 2 = 8 + 2$$
$$10 = 10$$

21. 5 faces, 10 vertices, 13 edges

$$v + f = e + 2$$
$$10 + 5 = 13 + 2$$
$$15 = 15$$

23. $v + f = e + 2$

$$8 + f = 15 + 2$$
$$f = 9 \text{ faces}$$

THINK ABOUT IT, SECTION 10.4

1. Every planar graph is 4-colorable.

2. The chromatic number of a graph is the minimum number of colors needed to color a graph so that no edge connects vertices of the same color.

SECTION EXERCISES 10.4

1. Requires three colors:

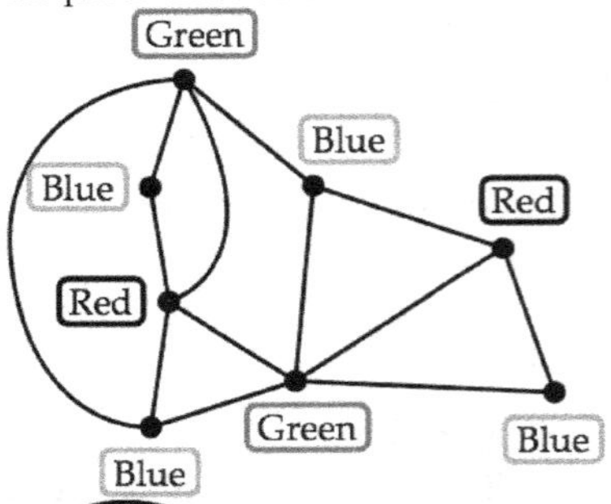

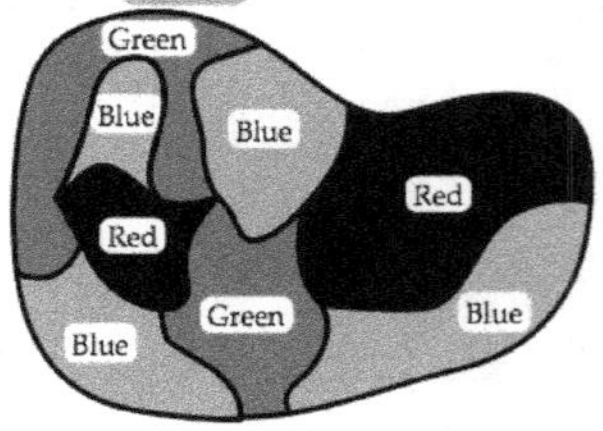

3. Requires three colors:

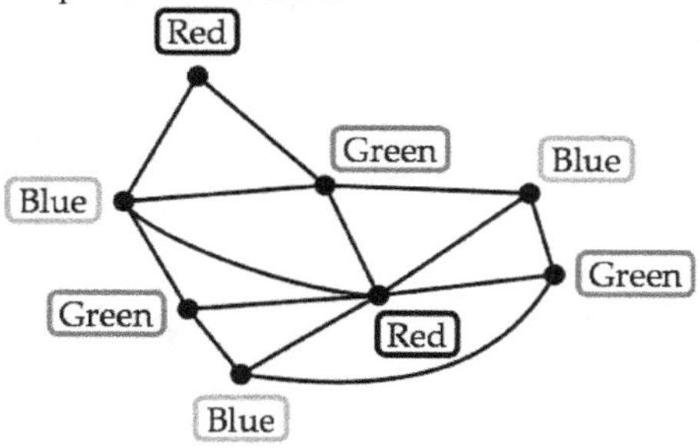

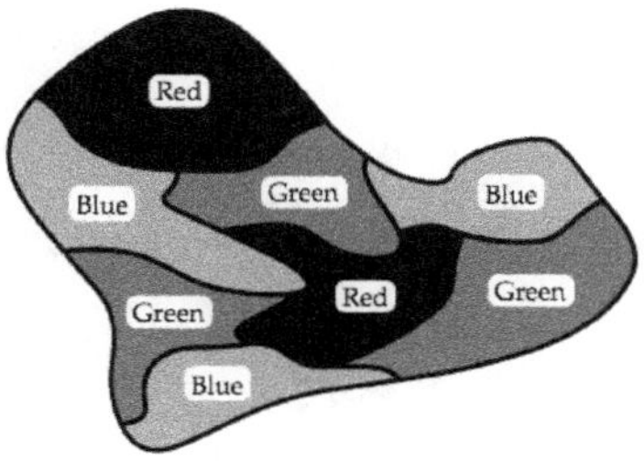

5. As a graph:

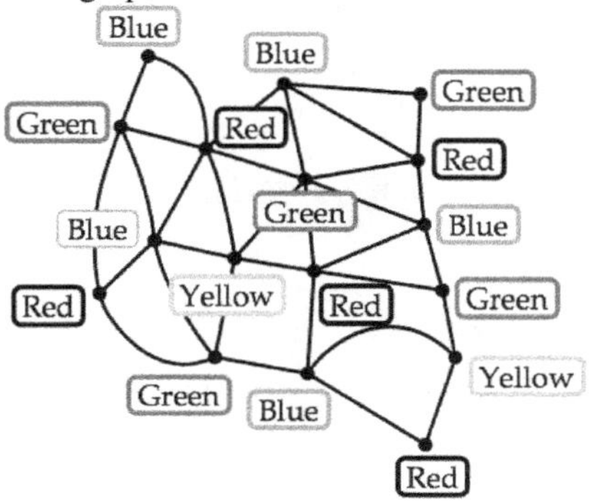

7.

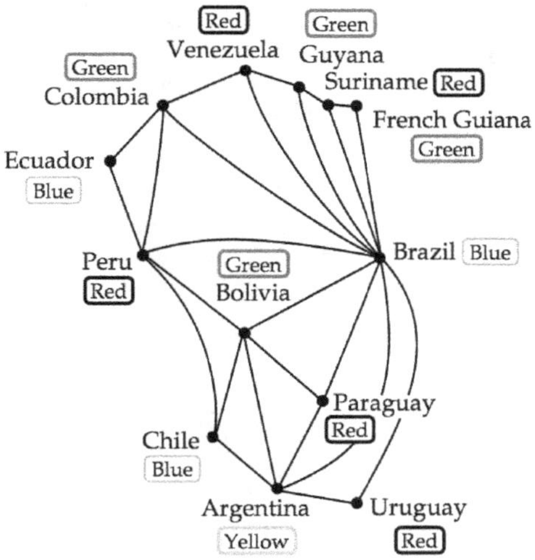

9. The graph is shown below:

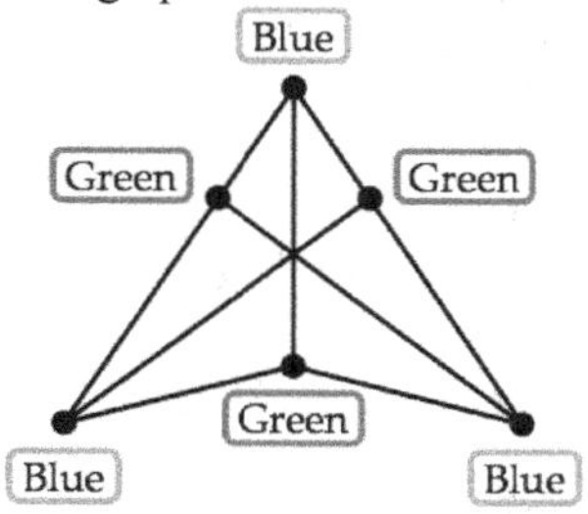

11. The graph is not 2-colorable because it has a circuit that consists of an odd number of vertices.

13. The graph is shown below:

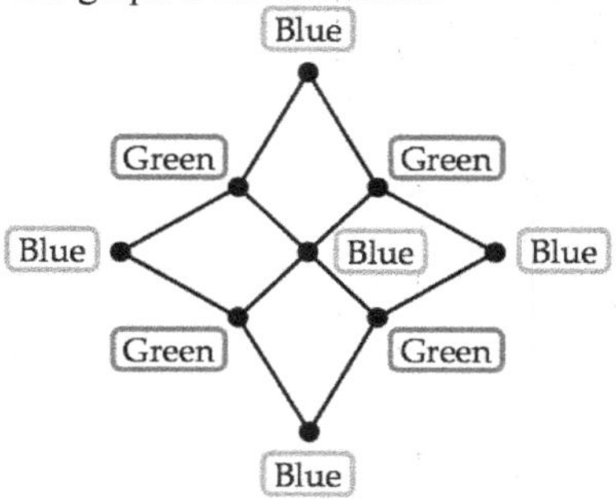

15. The graph is shown below:

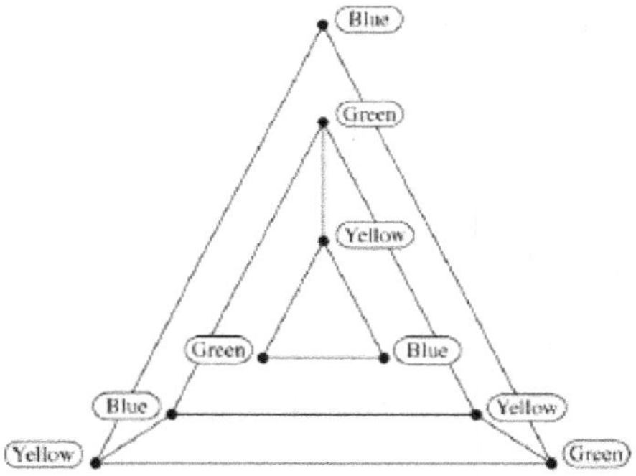

3

17. The graph is shown below:

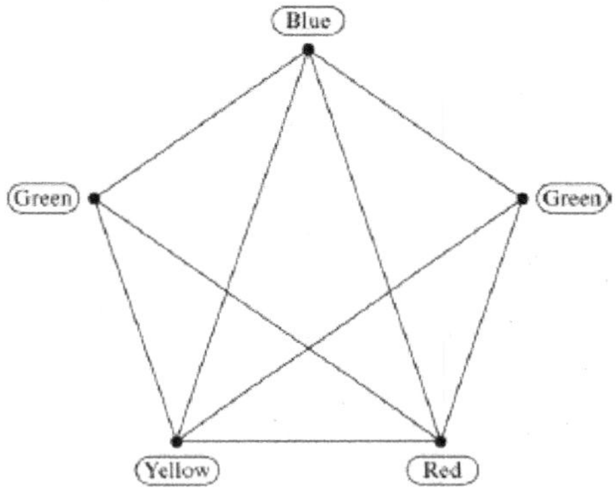

4

19. The graph is shown below:

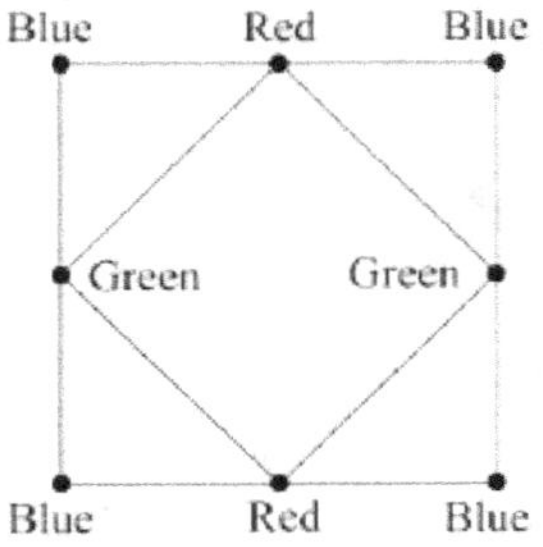

3

21. Represent the information as a graph:

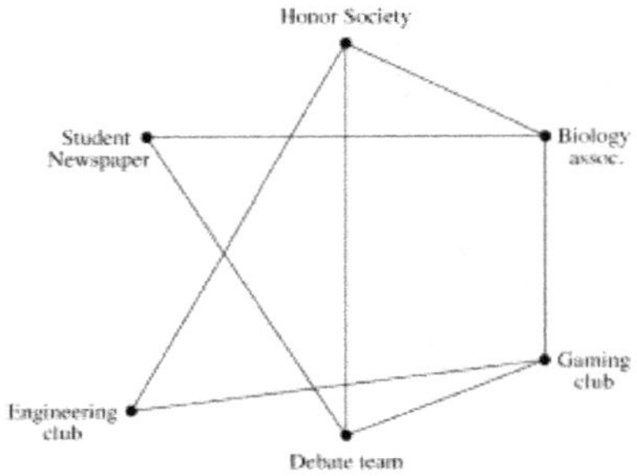

Two times slots are needed.

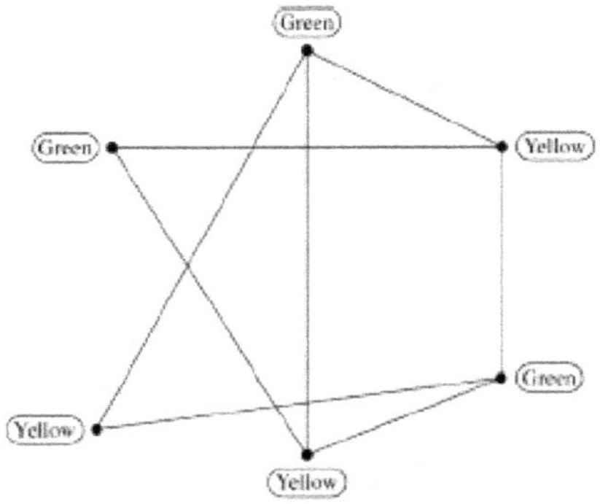

23. Represent the information as a graph:

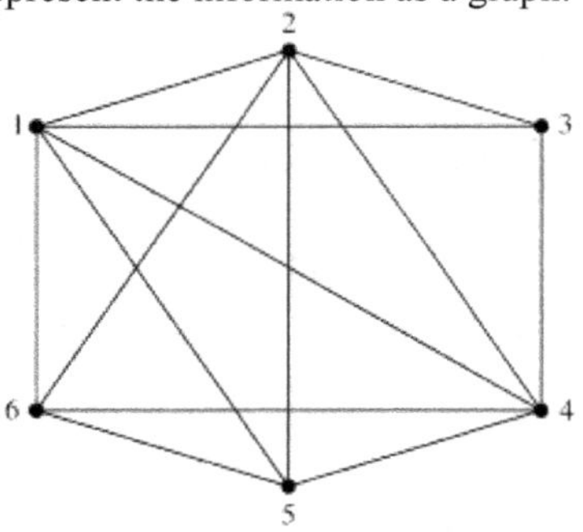

5 days are required.
Day 1: Group 1
Day 2: Group 2
Day 3: Groups 3 and 5
Day 4: Group 4
Day 5: Group 6

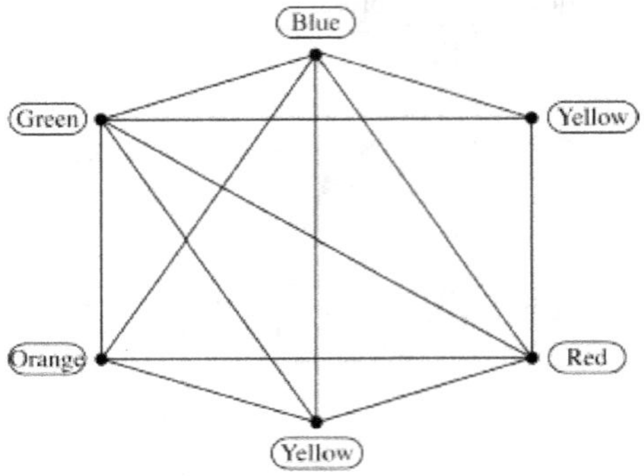

25. Represent the information as a graph:

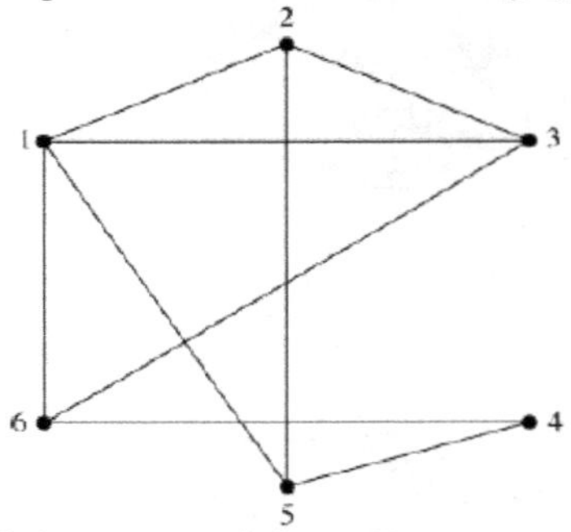

3 days are needed.

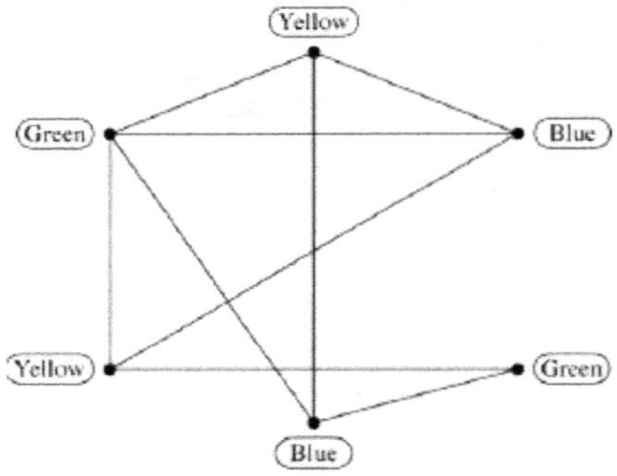

Day 1: Films 1 and 4
Day 2: Films 2 and 6
Day 3: Films 3 and 5

CHAPTER 10 REVIEW EXERCISES

1. a. 8
   b. 4
   c. All vertices have degree 4.
   d. Yes

2. a. 6
   b. 7
   c. 1, 1, 2, 2, 2, 2, 2
   d. No

3. As a graph:

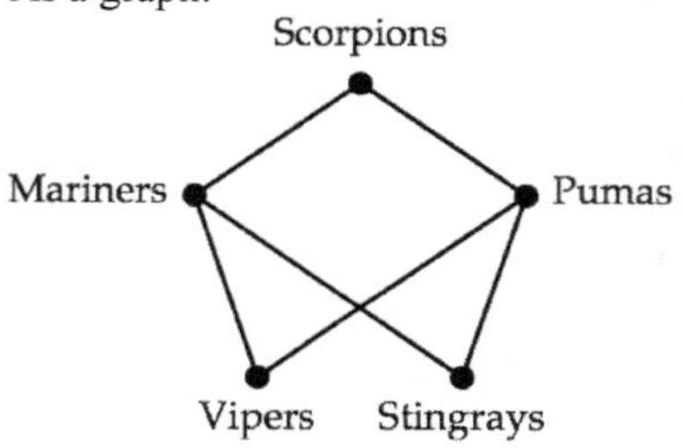

4. a. No
   b. 4
   c. 110, 405
   d. 105

5. Equivalent

6. Equivalent

7. a. E-A-B-C-D-B-E-C-A-D is an Euler walk.
   b. It is not possible to find an Euler circuit because there are vertices of odd degree.

8. a. It is not possible to find an Euler walk because there are more than 2 vertices of odd degree.
   b. It is not possible to find an Euler circuit because there are vertices of odd degree.

9. a. F-A-E-C-B-A-D-B-E-D-C-F is an Euler walk.
   b. F-A-E-C-B-A-D-B-E-D-C-F is an Euler circuit.

10. a. B-A-E-C-A-D-F-C-B-D-E is an Euler walk.
    b. It is not possible to find an Euler circuit because there are vertices of odd degree.

11. As a graph:

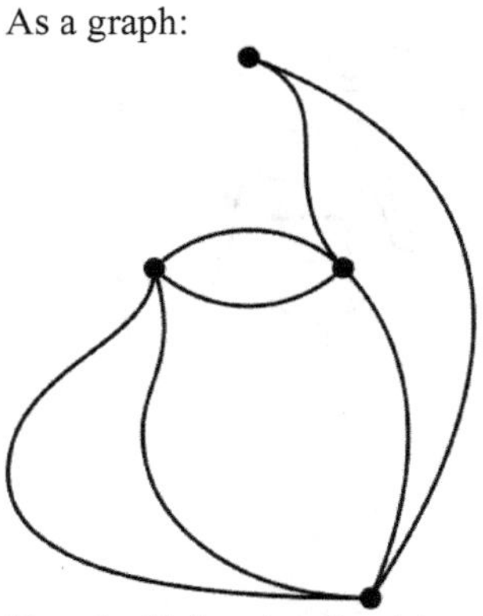

Yes. An Euler circuit exists.

12. Yes, it is possible to walk through each doorway exactly once, i.e. an Euler walk exists. No, it is not possible to do so and return to the starting point, i.e. an Euler circuit does not exist.

13. The graph is connected and has at least three vertices and no multiple edges, so the theorem applies. Every vertex has a degree of 3 or more, so the graph is Hamiltonian. A Hamiltonian circuit is A-B-C-E-D-A.

14. The graph is connected and has at least three vertices and no multiple edges, so the theorem applies. Every vertex has a degree of 4 or more, so the graph is Hamiltonian. A Hamiltonian circuit is A-D-F-B-C-E-A.

15. As a graph:

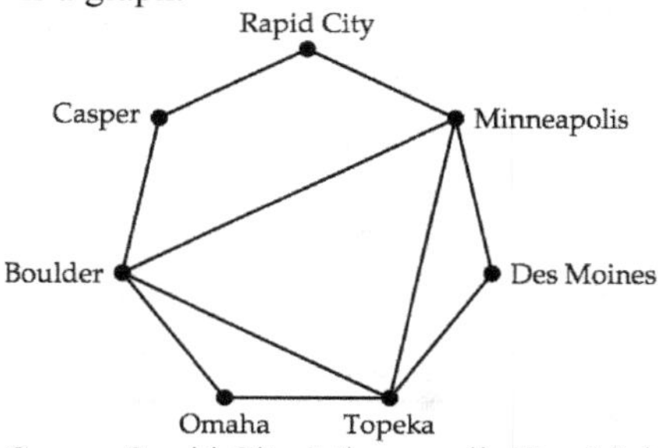

Casper-Rapid City-Minneapolis-Des Moines-Topeka-Omaha-Boulder-Casper

16. Casper-Boulder-Topeka-Minneapolis-Boulder-Omaha-Topeka-Des Moines-Minneapolis-Rapid City-Casper

17. A-D-F-E-B-C-A

18. A-B-E-C-D-A

19. A-D-F-E-C-B-A

20. A-B-E-D-C-A

21. As a graph:

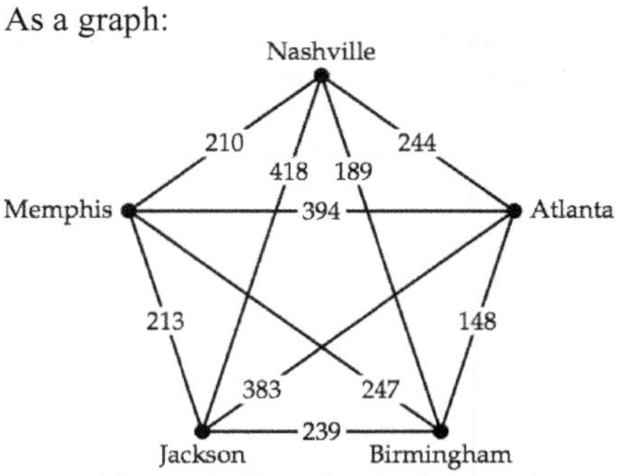

Memphis-Nashville-Birmingham-Atlanta-Jackson-Memphis

22. As a graph:

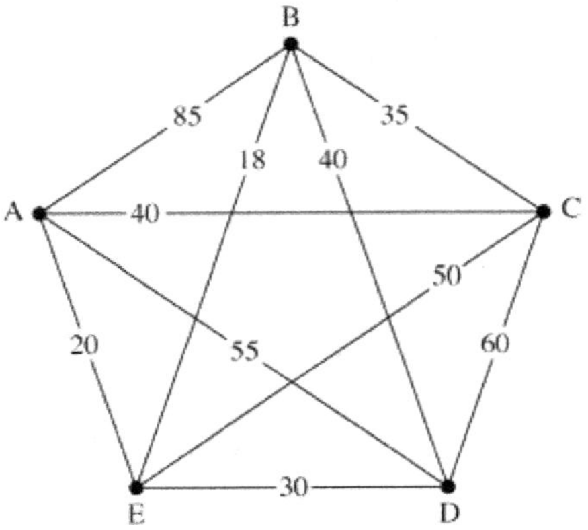

A-E-B-C-D-A

23. Redrawing:

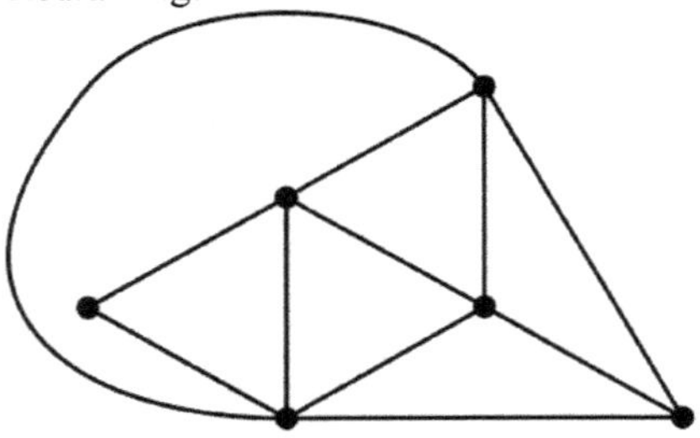

24. Redrawing:

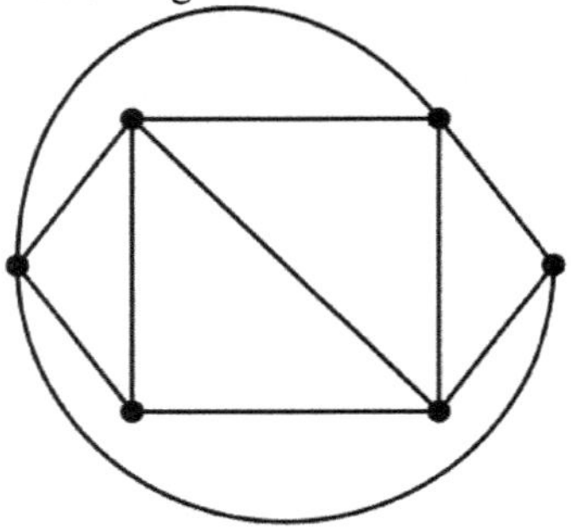

25. K_5 is a subgraph of the graph, so it is not planar.

26. The graph can be contracted into the Utilities Graph, so it is not planar.

27. 5 vertices, 8 edges, 5 faces

$v + f = e + 2$

$5 + 5 = 8 + 2$

$10 = 10$

28. 14 vertices, 16 edges, 4 faces

$v + f = e + 2$

$14 + 4 = 16 + 2$

$18 = 18$

29. As a graph:

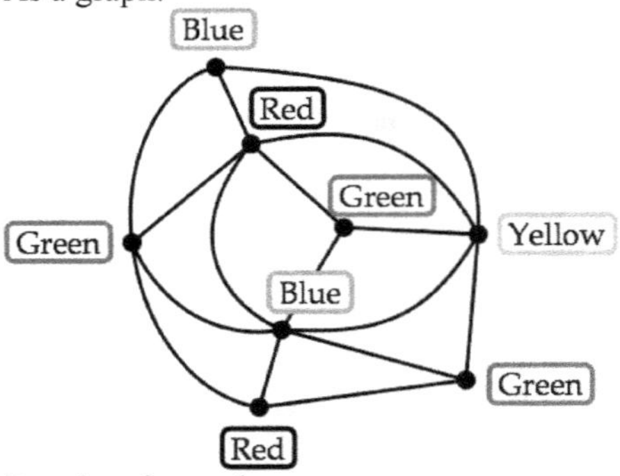

Requires four colors.

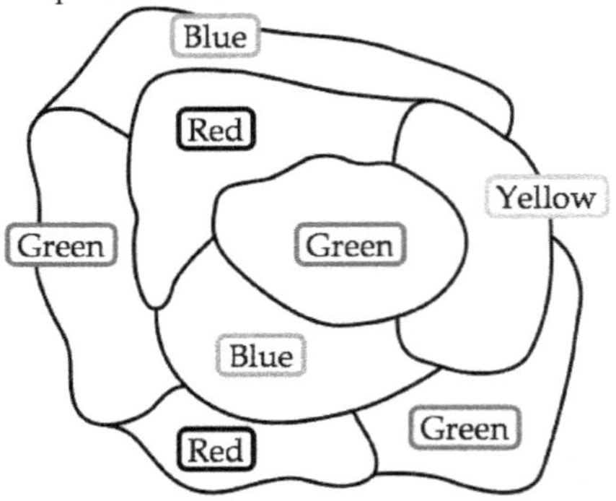

30. As a graph:

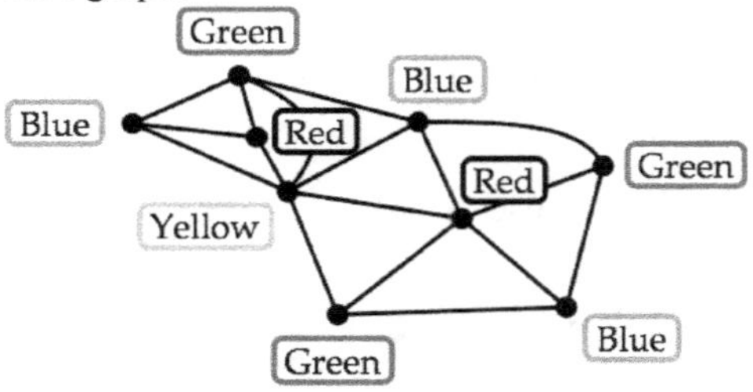

Requires four colors.

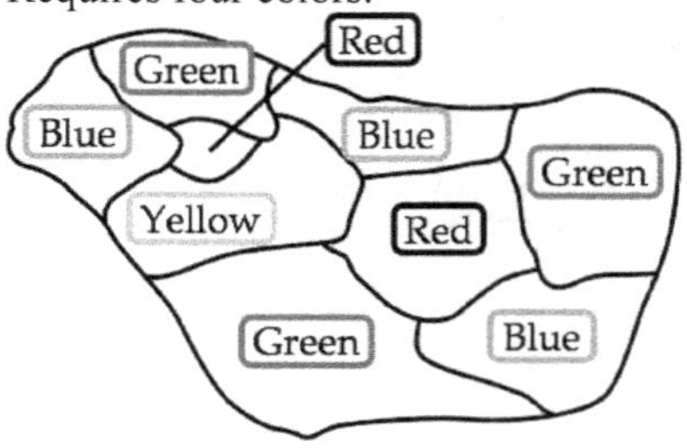

31. The graph has no circuits consisting of an odd number of vertices, so it is 2-colorable.

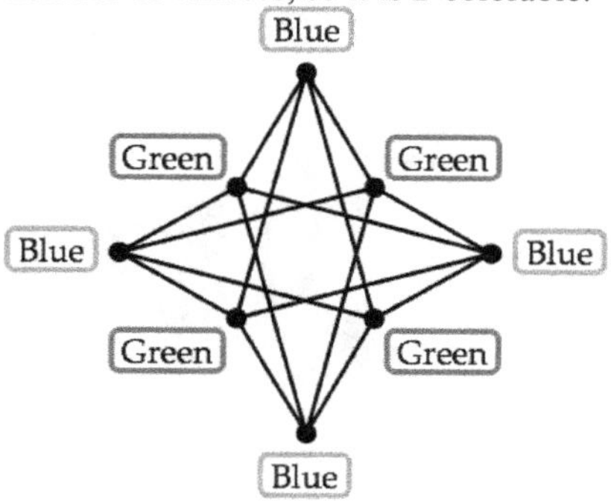

32. The graph has a circuit that consists of an odd number of vertices, so it is not 2-colorable.

33. The chromatic number is 3.

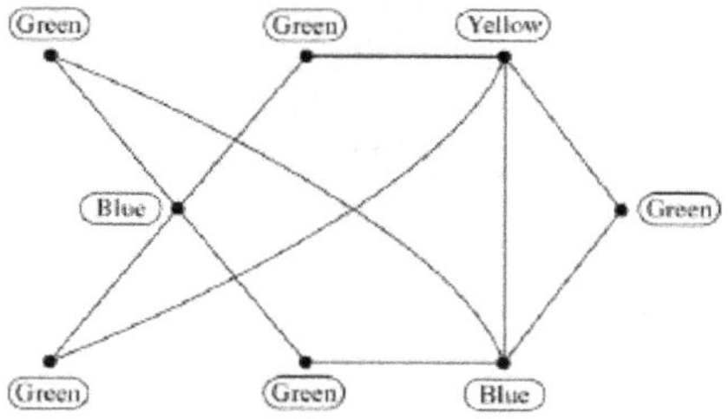

34. The chromatic number is 5.

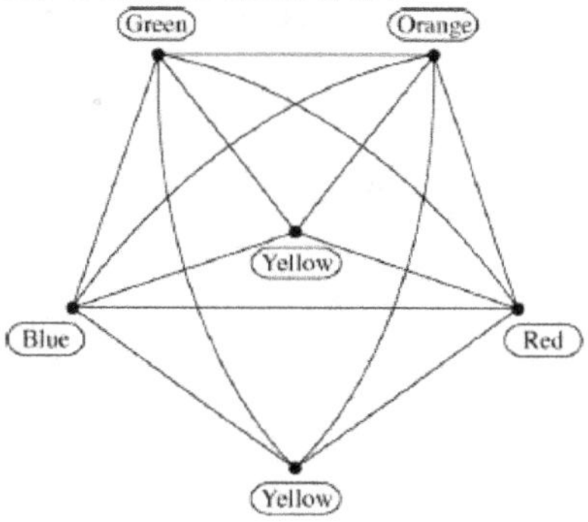

35. As a graph:

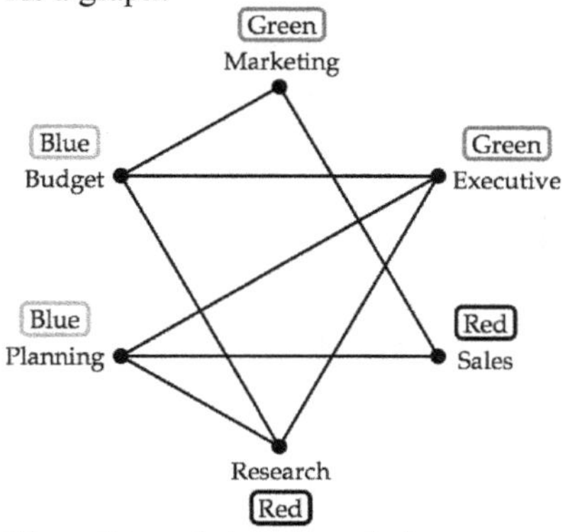

Three times slots are needed.
Time 1: Budget and Planning
Time 2: Marketing and Executive
Time 3: Sales and Research

Chapter 11: Geometry

THINK ABOUT IT, SECTION 11.1

1. $\dfrac{1 \text{ foot}}{12 \text{ inches}}$

2. 4 quarts in 1 gallon

3. 1000 ml = 1 L

4. 0.001 kg = 1 g

5. Approximately 4 liters in one gallon

6. 1 yard is less than 1 meter

7. Two pounds is less than 1 kilogram.

SECTION EXERCISES 11.1

1. $$\begin{aligned} 6 \text{ ft} &= 6 \cancel{\text{ft}} \times \frac{12 \text{ in.}}{1 \cancel{\text{ft}}} \\ &= 6 \times 12 \text{ in.} \\ &= 72 \text{ in.} \end{aligned}$$

3. $$\begin{aligned} 5 \text{ yd} &= 5 \cancel{\text{yd}} \times \frac{3 \cancel{\text{ft}}}{1 \cancel{\text{yd}}} \times \frac{12 \text{ in.}}{1 \cancel{\text{ft}}} \\ &= 5 \times 3 \times 12 \text{ in.} \\ &= 180 \text{ in.} \end{aligned}$$

5. $$\begin{aligned} 64 \text{ oz} &= 64 \cancel{\text{oz}} \times \frac{1 \text{ lb}}{16 \cancel{\text{oz}}} \\ &= \frac{64 \text{ lb}}{16} \\ &= 4 \text{ lb} \end{aligned}$$

7. $$\begin{aligned} 1\tfrac{1}{2} \text{ lb} &= \tfrac{3}{2} \cancel{\text{lb}} \times \frac{16 \text{ oz}}{1 \cancel{\text{lb}}} \\ &= \frac{48 \text{ oz}}{2} \\ &= 24 \text{ oz} \end{aligned}$$

9. $$\begin{aligned} 2\tfrac{1}{2} \text{ c} &= \tfrac{5}{2} \cancel{\text{c}} \times \frac{8 \text{ fl oz}}{1 \cancel{\text{c}}} \\ &= \frac{40 \text{ ft oz}}{2} \\ &= 20 \text{ fl oz} \end{aligned}$$

11. $$\begin{aligned} 7 \text{ gal} &= 7 \cancel{\text{gal}} \times \frac{4 \cancel{\text{qt}}}{1 \cancel{\text{gal}}} \times \frac{2 \text{ pt}}{1 \cancel{\text{qt}}} \\ &= 7 \times 4 \times 2 \text{ pt} \\ &= 56 \text{ pt} \end{aligned}$$

13. 62 cm = 620 mm

15. 3.21 m = 321 cm

17. 7421 g = 7.421 kg

19. 0.45 g = 4.5 dg

21. 7.5 ml = 0.0075 L

23. 0.435 L = 435 ml = 435 cm^3

25. $145 \text{ lb} \approx 145 \cancel{\text{lb}} \times \dfrac{1 \text{ kg}}{2.2 \cancel{\text{lb}}} = \dfrac{145 \text{ kg}}{2.2} \approx 65.91 \text{ kg}$

 145 lb ≈ 65.91 kg

27. $$\begin{aligned} \frac{30 \text{ mi}}{\text{h}} &\approx \frac{30 \cancel{\text{mi}}}{\text{h}} \times \frac{1.61 \text{ km}}{1 \cancel{\text{mi}}} \\ &= \frac{30 \times 1.61 \text{ km}}{\text{h}} \\ &= 48.3 \text{ km/h} \end{aligned}$$

 30 mi/h ≈ 48.3 km/h

29. $$\begin{aligned} \frac{\$32.99}{\text{gal}} &\approx \frac{\$32.99}{\cancel{\text{gal}}} \times \frac{1 \cancel{\text{gal}}}{3.79 \text{ L}} \\ &= \frac{\$32.99}{3.79 \text{ L}} \approx \frac{\$8.704485}{\text{L}} \\ &\approx \frac{\$8.70}{\text{L}} \end{aligned}$$

 \$32.99/gal ≈ \$8.70/L

31. $$\begin{aligned} 35 \text{ mm} &= 3.5 \text{ cm} \\ &= 3.5 \cancel{\text{cm}} \times \frac{1 \text{ in.}}{2.54 \cancel{\text{cm}}} \\ &= \frac{3.5 \text{ in.}}{2.54} \approx 1.37795 \text{ in.} \\ &\approx 1.38 \text{ in.} \end{aligned}$$

 35 mm ≈ 1.38 in.

33. $$\begin{aligned} \frac{\$10}{\text{kg}} &\approx \frac{\$10}{\cancel{\text{kg}}} \times \frac{1 \cancel{\text{kg}}}{2.2 \text{ lb}} \\ &= \frac{\$10}{2.2 \text{ lb}} \approx \frac{\$4.545454}{\text{lb}} \\ &\approx \frac{\$4.55}{\text{lb}} \end{aligned}$$

 \$10/kg ≈ \$4.55/lb

Think About It, Section 11.2

1. a. ray
   b. line segment
   c. line

2. a. $\angle a = \angle c$

   b. $\angle d = \angle b$

   c. $m\angle c + m\angle D = 180°$

   d. $\angle a$ and $\angle c$ are called vertical (or opposite) angles

3. 90°

4. greater than

5. acute

6. 180°

Section Exercises 11.2

1. The three names for the angle are $\angle O$, $\angle AOB$, and $\angle BOA$.

3. Solving:
   $$x + 62° = 90°$$
   $$x = 28°$$
   A 28° angle

5. Solving:
   $$x + 162° = 180°$$
   $$x = 18°$$
   An 18° angle

7. An acute angle, since complementary angles are two angles whose sum is 90°.

9. An obtuse angle, since the sum of supplementary angles are 180°, and an acute angle is less than 90°.

11. Solving:
   $$AB + BC + CD = AD$$
   $$12 + BC + 9 = 35$$
   $$BC = 14 \text{ cm}$$

13. Solving:
   $$QR + RS = QS$$
   $$QR + 3QR = QS$$
   $$7 + 21 = QS$$
   $$QS = 28 \text{ ft}$$

15. Solving:
   $$m\angle LOM + m\angle MON = m\angle LON$$
   $$53° + m\angle MON = 139°$$
   $$m\angle MON = 86°$$

17. Solving:
   $$x + 74° = 145°$$
   $$x = 71°$$

19. Solving:
   $$x + 2x = 90°$$
   $$3x = 90°$$
   $$x = 30°$$

21. Solving:
   $$a + 53° = 180°$$
   $$a = 127°$$

23. Solving:
   $$a + 76° + 168° = 360°$$
   $$a = 116°$$

25. Solving:
   $$3x + 4x + 2x = 180°$$
   $$9x = 180°$$
   $$x = 20°$$

27. Solving:
   $$5x + x + 20° + 2x = 180°$$
   $$8x = 160°$$
   $$x = 20°$$

29. $\angle a$ is complementary to $\angle b$'s supplementary angle.
   $$m\angle a + m\angle x = 90°$$
   $$51° + m\angle x = 90°$$
   $$m\angle x = 39°$$
   $$m\angle b + m\angle x = 180°$$
   $$m\angle b + 39° = 180°$$
   $$m\angle b = 141°$$

31. Solving:
   $$x + 74° = 180°$$
   $$x = 106°$$

33. The two angles are vertical angles, so their measures are equal.
   $$5x = 3x + 22°$$
   $$2x = 22°$$
   $$x = 11°$$

35. $\angle a$ is a corresponding angle with the 38° angle.

   Therefore, $m\angle a = 38°$.
   $$m\angle a + m\angle b = 180°$$
   $$38° + m\angle b = 180°$$
   $$m\angle b = 142°$$

37. $\angle a$ is an alternate interior angle with the 47° angle. Therefore, $m\angle a = 47°$.

$$\begin{aligned} m\angle a + m\angle b &= 180° \\ 47° + m\angle b &= 180° \\ m\angle b &= 133° \end{aligned}$$

39. Solving:

$$\begin{aligned} 5x + 4x &= 180° \\ 9x &= 180° \\ x &= 20° \end{aligned}$$

41. Solving:

$$\begin{aligned} 2x + x + 39° &= 180° \\ 3x &= 141° \\ x &= 47° \end{aligned}$$

43. The three lines form a triangle. The sum of the interior angles of the triangle is

$$\begin{aligned} m\angle b + (180° - m\angle a) + (180° - m\angle x) &= 180° \\ 70° + (180° - 95°) + (180° - m\angle x) &= 180° \\ 70° + 265° - m\angle x &= 180° \\ -m\angle x &= -155° \\ m\angle x &= 155° \end{aligned}$$

$\angle y$ and $\angle b$ are vertical angles:

$m\angle y = m\angle b = 70°$

45. $\angle a$ and $\angle y$ are vertical angles $m\angle a = \angle y = 45°$

The three lines form a triangle. The sum of the interior angles of the triangle is

$$\begin{aligned} m\angle a + 90° + (180° - m\angle b) &= 180° \\ 45° + 90° - m\angle b &= 0° \\ -m\angle b &= -135° \\ m\angle b &= 135° \end{aligned}$$

47. Solving:

$$\begin{aligned} 90° + 30° + x &= 180° \\ x &= 60° \end{aligned}$$

49. Solving:

$$\begin{aligned} 42° + 103° + x &= 180° \\ x &= 35° \end{aligned}$$

51. False, the sum of the measures of the interior angles of a triangle must be 180°. Two obtuse angles would be greater than 180°.

53. True, an obtuse angle is greater than 90°, and the sum of the measures of the interior angles of the triangle is 180°. Therefore, the remaining angles in the triangle would be less than 90°.

THINK ABOUT IT, SECTION 11.3

1. A rectangle has four 90° angles.

2. A square has four 90° angles and four equal sides. A rhombus has four equal sides.

3. The area of a square that measures 1 meter on each side is 1 m^2.

4. Twenty squares, each 1 foot on each side, are required to cover a rectangle that measures 20 ft^2.

5. False. $\pi \approx \frac{22}{7}$

6. a and c
The units in **a** are square inches.
The units in **c** are square meters.

7. b and d
The units in **b** are centimeters.
The units in **d** are feet.

SECTION EXERCISES 11.3

1. a. Perimeter is measured in linear units, not square units.

   b. Area is measured in square units, not linear units.

3. a. $P = 2l + 2w = 2(5) + 2(10) = 30$ m

   b. $A = lw = (5)(10) = 50\ \text{m}^2$

5. a. $P = 4s = 4(10) = 40$ km

   b. $A = s^2 = (10)^2 = 100\ \text{km}^2$

7. a. $P = 2(b) + 2(s) = 2(12) + 2(8) = 40$ ft

   b. $A = bh = (12)(6) = 72\ \text{ft}^2$

9. a. $C = 2\pi r = 2\pi(4) = 8\pi\ \text{cm} \approx 25.13$ cm

   b. $A = \pi r^2 = \pi(4)^2 = 16\pi\ \text{cm}^2 \approx 50.27\ \text{cm}^2$

11. a. $C = 2\pi r = 2\pi(5.5) = 11\pi\ \text{mi} \approx 34.56$ mi

    b. $A = \pi r^2 = \pi(5.5)^2 = 30.25\pi\ \text{mi}^2 \approx 95.03\ \text{mi}^2$

13. a. $C = \pi d = \pi(17) = 17\pi\ \text{ft} \approx 53.41$ ft

    b. $r = \frac{d}{2} = \frac{17}{2} = 8.5$ ft
    $A = \pi r^2 = \pi(8.5)^2 = 72.25\pi\ \text{ft}^2 \approx 226.98\ \text{ft}^2$

15. $P = 5s = 5(4) = 20$ in.

17. $P = 2b + 2s = 2(3) + 2(2) = 10$ mi

19. $P = 2l + 2w = 2(68) + 2(42) = 220$ in.
 15 ft = (15)(12) = 180 in.
 1 package is not enough.
 2 packages = 2(180) = 360 in.
 Two packages are needed.

21. The perimeter of the square equals $4s$. the circumference of the circle equals πs. Since $4 > \pi$, the perimeter of the square is greater.

23. $A = s^2 = (12)^2 = 144 \text{ m}^2$

25. $P = 4s = 36$ in.
 $s = 9$ in.

27. $A = lw$
 $(l)(8) = 312 \text{ ft}^2$
 $l = 39$ ft

29. Comparing:
 $10 \times 5 = 50$
 $10 \times 10 = 100$
 $10 \times 15 = 150$
 $10 \times 20 = 200$
 The smallest unit that will provide 175 ft^2 of floor space is 10 feet by 20 feet.

31. $A = \frac{1}{2}bh = \frac{1}{2}(21)(13) = 136.5 \text{ ft}^2$

33. $A = \frac{1}{2}h(b_1 + b_2) = \frac{1}{2}(10)(23 + 9) = 160 \text{ km}^2$

35. $A = \frac{1}{2}h(b_1 + b_2) = \frac{1}{2}(10)(10 + 12) = 110 \text{ ft}^2$
 110 ft^2 ÷ 55 ft^2 per quart = 2 quarts

37. $A = lw$
 Wall 1 = (10)(8) = 80 ft^2
 Wall 2 = (12)(8) = 96 ft^2
 80 ft^2 + 96 ft^2 = 176 ft^2 total area
 176 ft^2 ÷ 40 ft^2 per roll = 4.4, which is 5 rolls
 5 × \$96 = \$480

39. $A = lw = (8)(9) = 72 \text{ ft}^2$
 $A = lw = (12)(12) = 144 \text{ ft}^2$
 72 + 144 = 216 ft^2 total area
 216 ft^2 ÷ 9 ft^2 per yard^2 = 24 yd^2
 24 yd^2 × \$38 = \$912

41. Length of total region = 25 + 4 = 29 m
 Width of total region = 15 + 4 = 19 m
 Total Area = lw = (29)(19) = 551 m^2
 Area of Plot = lw = (25)(15) = 375 m^2
 Area of Walkway = 551 – 375 = 176 m^2

43. $C = \pi d = \pi(4.2) \approx 13.19$ ft

45. $C = \pi d = \pi(16) \approx 50.265$ in.
 50.265 in. per revolution × 240 revolutions per minute = 12,064 in. per minute

47. $C = \pi d = \pi(16) \approx 50.265$ in.
 50.265 in × 15 revolutions ≈ 753.98 in.
 753.98 in. ÷ 12 ≈ 62.83 ft

49. $A = \pi r^2 = \pi(50)^2 = 2{,}500\pi \text{ ft}^2$

51. $A = \pi r^2$
 Area of Small Pizza = $\pi(6)^2 \approx 113.10 \text{ in}^2$
 Area of Large Pizza = $\pi(12)^2 \approx 452.39 \text{ in}^2$
 452.39 – 113.10 = 339.29 in^2
 2(113.10) = 226.10 in^2, the area of the large pizza is more than twice that of the small pizza; it is four times the area of the small pizza.

53. $A = \frac{1}{2}h(b_1 + b_2)$

$$A = \frac{1}{2}(2.75)(9 + 7.8) = 23.1$$
$$A = \frac{1}{2}(2.75)(10.3 + 9) = 26.5375$$
$$A = \frac{1}{2}(2.75)(12 + 10.3) = 30.6625$$
$$A = \frac{1}{2}(2.75)(11.3 + 12) = 32.0375$$
$$A = \frac{1}{2}(2.75)(11.1 + 11.3) = 30.8$$
$$A = \frac{1}{2}(2.75)(9.8 + 11.1) = 28.7375$$
$$A = \frac{1}{2}(2.75)(9.2 + 9.8) = 26.125$$
$$A = \frac{1}{2}(2.75)(8.4 + 9.2) = 24.2$$

 A total = sum of all totals = 222.2 mi^2

55. Length of Rectangle = $4r$
 Width of Rectangle = $2r$
 Area of Rectangle = $(4r)(2r) = 8r^2$
 Area of Circles = $2\pi r^2$
 Area of shaded region = $8r^2 - 2\pi r^2$

57. $A = lw$
 Double the length and the width.
 $A = (2l)(2w) = 4(lw)$
 The area is four times that of the original rectangle.

THINK ABOUT IT, SECTION 11.4

1. Volume is a measure of capacity such as number of liters of milk in a container. Surface area is a measure of how much material is required to make the container.

2. The units of the volume is m^3.

3. The units of the surface area is in^2.

4. The shape of the base of a right circular cylinder is a circle.

SECTION EXERCISES 11.4

1. $V = lwh = (14)(10)(6) = 840 \text{ in}^3$

3. $V = \frac{1}{3}s^2h = \frac{1}{3}(3)^2(5) = 15 \text{ ft}^3$

5. Solving:
$V = \frac{1}{2}d = \frac{1}{2}(3) = \frac{3}{2}$
$V = \frac{4}{3}\pi r^3 = \frac{4}{3}\pi\left(\frac{3}{2}\right)^3 = \left(\frac{4}{3}\right)\left(\frac{27}{8}\right)\pi$
$V = 4.5\pi \text{ cm}^3 \approx 14.14 \text{ cm}^3$

7. Solving:
$l = 4 \quad w = 5 \quad h = 3$
$S = 2lw + 2lh + 2\,wh$
$S = 2(4)(5) + 2(4)(3) + 2(5)(3)$
$S = 94 \text{ m}^2$

9. Solving:
$S = 4 \quad l = 5$
$S = s^2 + 2sl$
$= 4^2 + 2(4)(5)$
$16 + 40$
$S = 56 \text{ m}^2$

11. Solving:
$r = 6 \quad h = 2$
$S = 2\pi r^2 + 2\pi rh$
$S = 2\pi(6)^2 + 2\pi(6)(2)$
$S = 96\pi \text{ in}^2 \approx 301.59 \text{ in}^2$

13. $V = lwh = (6.8)(2.5)(2) = 34 \text{ m}^3$

15. $V = s^3 = (2.5)^3 = 15.625 \text{ in}^3$

17. Solving:
$V = \frac{4}{3}\pi r^3 \quad r = 3$
$V = \frac{4}{3}\pi(3)^3 = 36\pi \text{ ft}^3$

19. Solving:
$d = 24 \quad r = \frac{1}{2}(24) = 12 \quad h = 18$
$V = \pi r^2 h$
$V = \pi(12)^2(18) = 2592\pi \approx 8143.01 \text{ cm}^3$

21. Solving:
$r = 5 \quad h = 9$
$V = \frac{1}{3}\pi r^2 h$
$V = \frac{1}{3}\pi(5)^2(9) = 75\pi \text{ in}^3$

23. Solving:
$s = 6 \quad h = 10$
$V = \frac{1}{3}s^2h = \frac{1}{3}(6)^2(10) = 120 \text{ in}^3$

25. Since the length of the side of the cube equals the radius of the sphere, then $s = r$.
Volume of cube $= s^3$
$= r^3$
Volume of sphere $= \frac{4}{3}\pi r^3$
Sphere has a greater volume, because $\frac{4}{3}\pi > 1$.

27. Solving:
$C = 3.5 \quad h = 8$
If $C = 2\pi r$, then $r = \frac{C}{2\pi} = \frac{3.5}{2\pi} \approx 0.557$
$V = \pi r^2 h = \pi(0.557)^2(8) \approx 7.80 \text{ ft}^3$

29. $l = 1000 \quad w = 110 \quad h = 43$
$V = lwh = (1000)(110)(43) = 4{,}730{,}000 \text{ ft}^3$
$(4{,}730{,}000 \text{ ft}^3)(7.48 \text{ gal/ft}^3) = 35{,}380{,}400 \text{ gal}$

31. Solving:
$s = 3.4$
$S = 6s^2 = 6\,(3.4)^2 = 69.36 \text{ m}^2$

33. Solving:
$r = \frac{1}{2}(15) = 7.5$
$S = 4\pi r^2 = 4\pi(7.5)^2 = 225\pi \text{ cm}^2$

35. Solving:
$r = 4 \quad h = 12$
$S = 2\pi r^2 + 2\pi rh = 2\pi(4)^2 + 2\pi(4)(12)$
$S = 32\pi + 96\pi = 128\pi \approx 402.12 \text{ in}^2$

37. Solving:
$r = 1.5 \quad l = 2.5$
$S = \pi r^2 + \pi rl$
$S = \pi(1.5)^2 + \pi(1.5)(2.5)$
$S = 2.25\pi + 3.75\pi = 6\pi \text{ ft}^2$

39. Solving:
$s = 9 \quad l = 12$
$S = s^2 + 2sl = (9)^2 + 2(9)(12)$
$S = 297 \text{ in}^2$

41. Solving:
$l = 7 \quad h = 3 \quad V = 52.5$
$V = lwh$
$w = \dfrac{V}{lh}$
$w = \dfrac{52.5}{7(3)} = \dfrac{52.5}{21} = 2.5 \text{ ft}$

43. Solving:
$r = 12 \quad h = 30$
$S = 2\pi r^2 + 2\pi rh$
$S = 2\pi(12)^2 + 2\pi(12)(30)$
$S = 1008\pi \approx 3166.73 \text{ ft}^2$
Each can covers 300 ft^2.
$\dfrac{S}{300} = \dfrac{3166.73}{300} \approx 10.56$
11 cans are needed.

45. Solving:
Pyramid:
$S = s^2 + 2sl = (5)^2 + 2(5)(8) = 105 \text{ cm}^2$
Cone:
$r = 2.5$
$S = \pi r^2 + \pi rl$
$S = \pi(2.5)^2 + \pi(2.5)(8) = 26.25\pi \approx 82.47 \text{ cm}^2$
Difference $= 105 - 82.47 = 22.53 \text{ cm}^2$

47. Break figure into 2 rectangular prisms.
Prism 1 = $1.5 \times 1.5 \times 2$
$V_1 = (1.5)(1.5)(2) = 4.5$
Prism 2 = $0.5 \times 0.5 \times 2$
$V_2 = (0.5)(0.5)(2) = 0.5$
$V_{\text{total}} = V_1 + V_2$
$V_{\text{total}} = 4.5 + 0.5 = 5 \text{ m}^3$

49. $V_{\text{total}} = V_1 + V_2$
$V_1 = \pi(3)^2(2) = 18\pi$
$V_2 = \pi(1)^2(4) = 4\pi$
$V_{\text{total}} = V_1 + V_2 = 18\pi + 4\pi$
$V_{\text{total}} = 22\pi \approx 69.12 \text{ in}^3$

51. $V_{\text{total}} = V_{\text{prism}} + V_{\text{pyramid}}$
Rectangular Prism:
$V_{\text{prism}} = lwh = (8)(8)(2) = 128 \text{ in}^3$
Square Pyramid:
$h = 5 - 2 = 3$ in.
$V_{\text{pyramid}} = \dfrac{1}{3}s^2h = \dfrac{1}{3}(8)^2(3) = 64 \text{ in}^3$
$V_{\text{total}} = 128 + 64 = 192 \text{ in}^3$

53. Solving:
$S_{\text{total}} = S_{\text{bottom}} + 4S_{\text{rect.face}} + 4S_{\text{triangle face}}$
$S_{\text{bottom}} = (8)(8) = 64$
$S_{\text{rect.face}} = (2)(8) = 16$
$S_{\text{triangle face}} = \dfrac{1}{2}(8)(5) = 20$
$S_{\text{total}} = 64 + 4(16) + 4(20) = 208 \text{ in}^2$

55. Solving:
$S_{\text{total}} = S_{\text{cylinder}} + S_{\text{prism}} - 2S_{\text{circle}}$
$S_{\text{cylinder}} = 2\pi(1)^2 + 2\pi(1)(2) = 6\pi$
$S_{\text{prism}} = 2(8)(2) + 2(8)(2) + 2(8)(8) = 192$
$S_{\text{circle}} = \pi(1)^2 = \pi$
$S_{\text{total}} = (6\pi + 192) - 2(\pi) = 204.57 \text{ cm}^2$

57. Solving:
$V = 6(2.5)(5) + 4(1)(5) = 95 \text{ m}^3$
$V = (95 \text{ m}^3)\left(\dfrac{1000 \text{ L}}{1 \text{ m}^3}\right) = 95{,}000 \text{ L}$

59. Solving:
$V_s = \dfrac{4}{3}r_s^3$
$V_l = \dfrac{4}{3}r_l^3$
$r_l = 3r_s$
$V_l = \dfrac{4}{3}(3r_s)^3 = 27V_s$
Value of large sphere is 27 times the value of the small sphere or \$4860.

THINK ABOUT IT, SECTION 11.5

1. If two triangles are similar, corresponding angles are equal.

2. The side corresponding to AB is DE.

3. Similar triangles may have the same area, but generally do not. Congruent triangles always have the same area.

4. Similar triangles are generally not congruent. Congruent triangles are always similar triangles.

5. The side-angle-side theorem for congruent triangles states if two sides and the included angle of one triangle are equal in measure to two sides and the included angle of a second triangle, the two triangles are congruent.

6. Yes, the triangle is a right triangle.
$9^2 + 12^2 = 15^2$

SECTION EXERCISES 11.5

1. $\dfrac{7}{14} = \dfrac{1}{2}$

3. $\dfrac{6}{8} = \dfrac{3}{4}$

5. $\frac{5}{9}=\frac{4}{x}$
$5x=36$
$x=7.2\text{ cm}$

7. $\frac{3}{5}=\frac{2}{x}$
$3x=10$
$x\approx 3.3\text{ m}$

9. $\frac{4}{8}=\frac{x}{6}$
$8x=24$
$x=3\text{ m}$
$P=3+4+5=12\text{ m}$

11. $\frac{4}{12}=\frac{x}{15}$
$12x=60$
$x=5\text{ in.}$
$P=3+4+5=12\text{ in.}$

13. $\frac{15}{40}=\frac{x}{20}$
$40x=300$
$x=7.5\text{ cm}$
$A=\frac{1}{2}bh=\frac{1}{2}(15)(7.5)=56.3\text{ cm}^2$

15. $\frac{24}{8}=\frac{x}{6}$
$8x=144$
$x=18\text{ ft}$

17. $\frac{8}{4}=\frac{x}{8}$
$4x=64$
$x=16\text{ m}$

19. 5 ft 9 in. = 5.75 ft
$\frac{12}{30}=\frac{5.75}{x}$
$12x=172.5$
$x=14.375=14\frac{3}{8}\text{ ft}$

21. $\frac{8}{20}=\frac{6}{x}$
$8x=120$
$x=15\text{ m}$

23. Solving:
$\frac{BC}{AC}=\frac{CE-DE}{CE}$
$AC=3+3=6\text{ ft}$
$\frac{3}{6}=\frac{x-4}{x}$
$3x=6x-24$
$3x=24$
$x=8\text{ ft}$

25. $\frac{24}{12}=\frac{39-x}{x}$
$24x=468-12x$
$36x=468$
$x=13\text{ cm}$

27. $\frac{8}{20}=\frac{14}{x}$
$8x=280$
$x=35\text{ m}$

29. Yes, SAS Theorem.

31. Yes, SSS Theorem.

33. Yes, ASA Theorem.

35. No.

37. Yes, SAS Theorem.

39. No.

41. No.

43. Using the Pythagorean Theorem:
$a^2+b^2=c^2$
$3^2+4^2=c^2$
$25=c^2$
$5=c$
The side is 5 in.

45. Using the Pythagorean Theorem:
$a^2+b^2=c^2$
$5^2+7^2=c^2$
$74=c^2$
$8.6\approx c$
The side is 8.6 cm.

47. Using the Pythagorean Theorem:
$a^2+b^2=c^2$
$10^2+b^2=15^2$
$b^2=225-100=125$
$b\approx 11.2$
The side is 11.2 ft.

49. Using the Pythagorean Theorem:
$$a^2+b^2=c^2$$
$$4^2+b^2=6^2$$
$$b^2=36-16=20$$
$$b\approx 4.5$$
The side is 4.5 cm.

51. Using the Pythagorean Theorem:
$$a^2+b^2=c^2$$
$$9^2+9^2=c^2$$
$$162=c^2$$
$$12.7\approx c$$
The side is 12.7 yd.

53. Using the Pythagorean Theorem:
$$a^2+b^2=c^2$$
$$3^2+8^2=c^2$$
$$73=c^2$$
$$8.5\approx c$$
The holes are 8.5 cm apart.

55. Using the Pythagorean Theorem:
$$a^2+b^2=c^2$$
$$5^2+9^2=c^2$$
$$106=c^2$$
$$10.3\approx c$$
The hypotenuse has length 10.3 cm. Adding: $5 + 9 + 10.3 = 24.3$. The perimeter is 24.3 cm.

THINK ABOUT IT, SECTION 11.6

1. a. side b
   b. side a
   c. side a
   d. side b

2. $\sin\alpha=\frac{b}{c},\quad \cos\alpha=\frac{a}{c},\quad \tan\alpha=\frac{b}{a}$

3. $\tan^{-1}(1)=45°$

4. $\sin x=\theta$

SECTION EXERCISES 11.6

1. a. $\sin A=\frac{\text{opp}}{\text{hyp}}=\frac{a}{c}$

   b. $\sin B=\frac{\text{opp}}{\text{hyp}}=\frac{b}{c}$

   c. $\cos A=\frac{\text{adj}}{\text{hyp}}=\frac{b}{c}$

   d. $\cos B=\frac{\text{adj}}{\text{hyp}}=\frac{a}{c}$

   e. $\tan A=\frac{\text{opp}}{\text{adj}}=\frac{a}{b}$

   f. $\tan A=\frac{\text{opp}}{\text{adj}}=\frac{b}{a}$

3. Use the Pythagorean Theorem to find the unknown side:
$$a^2+b^2=c^2$$
$$12^2+5^2=c^2$$
$$169=c^2$$
$$13=c$$
$$\sin\theta=\frac{5}{13},\quad \cos\theta=\frac{12}{13},\quad \tan\theta=\frac{5}{12}$$

5. Use the Pythagorean Theorem to find the unknown side:
$$a^2+b^2=c^2$$
$$a^2+24^2=25^2$$
$$a^2=49$$
$$a=7$$
$$\sin\theta=\frac{24}{25},\quad \cos\theta=\frac{7}{25},\quad \tan\theta=\frac{24}{7}$$

7. Use the Pythagorean Theorem to find the unknown side:
$$a^2+b^2=c^2$$
$$8^2+7^2=c^2$$
$$113=c^2$$
$$\sqrt{113}=c$$
$$\sin\theta=\frac{8}{\sqrt{113}},\quad \cos\theta=\frac{7}{\sqrt{113}},\quad \tan\theta=\frac{8}{7}$$

9. Use the Pythagorean Theorem to find the unknown side:
$$a^2+b^2=c^2$$
$$a^2+40^2=80^2$$
$$a^2=4800$$
$$a=\sqrt{4800}=40\sqrt{3}$$
$$\sin\theta=\frac{1}{2},\quad \cos\theta=\frac{40\sqrt{3}}{80}=\frac{\sqrt{3}}{2},$$
$$\tan\theta=\frac{40}{40\sqrt{3}}=\frac{1}{\sqrt{3}}$$

11. 0.6820

13. 1.9970

15. 0.8453

17. 38.6°

19. 17.9°

21. 21.3°

23. 26.6°

25. 38.0°

27. 41.1°

29.

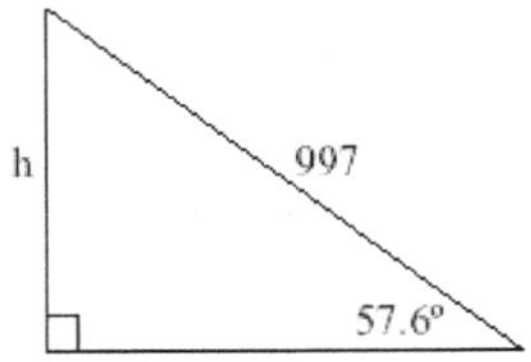

$$\sin 57.6° = \frac{h}{997}$$
$$h = 997 \sin 57.6° \approx 841.79$$
The balloon is 841.8 ft off the ground.

31.

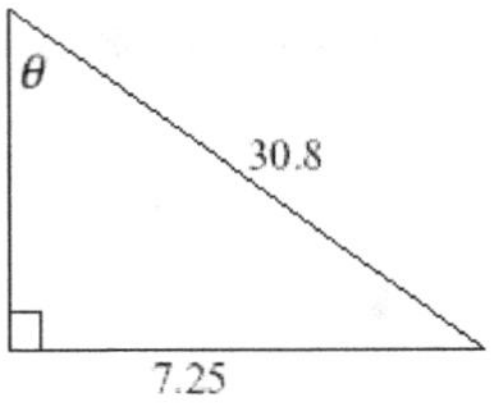

$$\sin \theta = \frac{7.25}{30.8}$$
$$\theta = \sin^{-1} \frac{7.25}{30.8} \approx 13.6$$
The angle is 13.6°.

33.

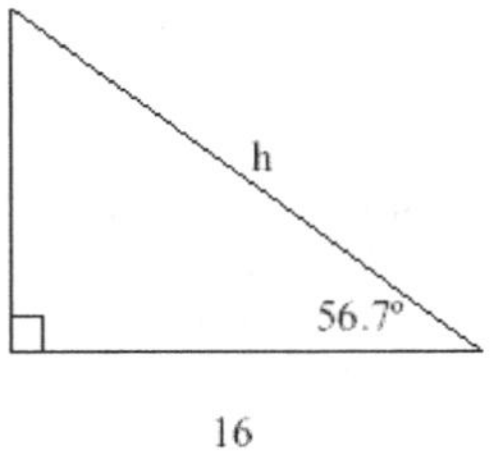

$$\cos 56.7° = \frac{16}{h}$$
$$h = \frac{16}{\cos 56.7°} \approx 29.14$$
The wire is 29.1 ft long.

35.

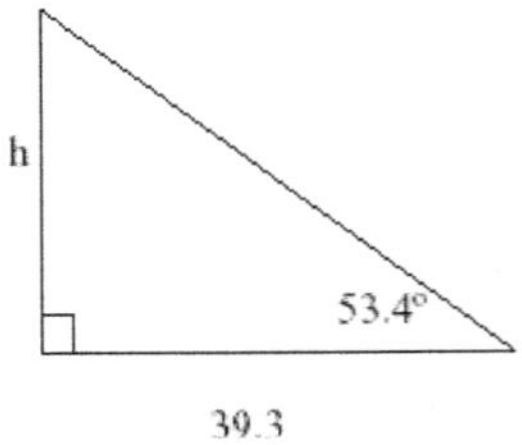

$$\tan 53.4° = \frac{h}{39.3}$$
$$h = 39.3 \tan 53.4° \approx 52.92$$
The height is 52.9 ft.

37.

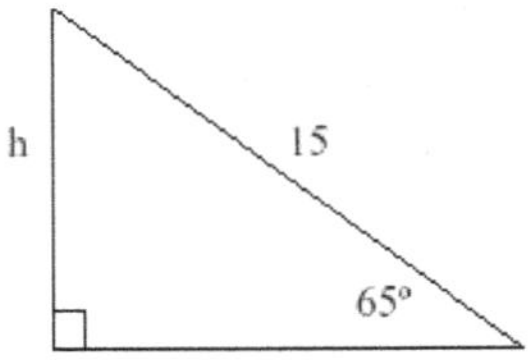

$$\sin 65 = \frac{h}{15}$$
$$h = 15 \sin 65° \approx 13.59$$
The ladder reaches 13.6 ft up the side.

39.

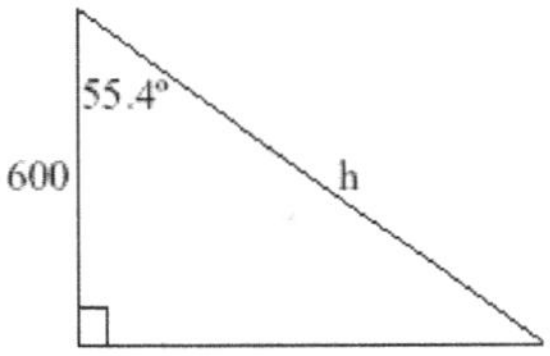

$$\cos 55.4° = \frac{600}{h}$$
$$h = \frac{600}{\cos 55.4°} \approx 1056.63$$
The wire is 1056.6 ft long.

41.

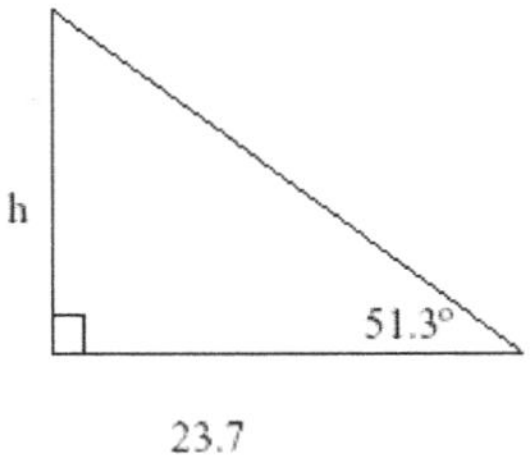

$$\tan 51.3° = \frac{h}{23.7}$$
$$h = 23.7 \tan 51.3° \approx 29.58$$
The tree is 29.6 yd tall.

THINK ABOUT IT, SECTION 11.7

1. a. Through a given point not on a given line, exactly one line can be drawn parallel to the given line.

   b. Through a given point not on a given line, there are at least two lines parallel to the given line.

   c. Through a given point not on a given line, there exist no lines parallel to the given line.

2. a. One

   b. One

   c. Three

3. A geodesic is a curve on a surface such that for any two points of the curve the portion of the curve between the points is the shortest path on the surface that joins these points.

4. Lobachevskian or hyperbolic geometry

5. Riemannian geometry

SECTION EXERCISES 11.7

1. Calculating:
$$m\angle A = 150°,\ \ m\angle B = 120°,\ \ m\angle C = 90°$$
$$S = (m\angle A + m\angle B + m\angle C - 180°)\left(\frac{\pi}{180°}\right)r^2$$
$$S = (150° + 120° + 90° - 180°)\left(\frac{\pi}{180°}\right)(1)^2$$
$$S = \pi \text{ units}^2$$

3. Using the formulas:
$$\begin{aligned} d_E(P,Q) &= \sqrt{(x_2 - x_1)^2 + (y_2 - y_1)^2} \\ &= \sqrt{(4-(-3))^2 + (1-1)^2} \\ &= \sqrt{49} = 7 \text{ blocks} \end{aligned}$$
$$\begin{aligned} d_C(P,Q) &= |x_2 - x_1| + |y_2 - y_1| \\ &= |4-(-3)| + |1-1| \\ &= |7| + |0| = 7 \text{ blocks} \end{aligned}$$

5. Using the formulas:
$$\begin{aligned} d_E(P,Q) &= \sqrt{(x_2 - x_1)^2 + (y_2 - y_1)^2} \\ &= \sqrt{(-3-2)^2 + (5-(-3))^2} \\ &= \sqrt{25+64} = \sqrt{89} \approx 9.4 \text{ blocks} \end{aligned}$$
$$\begin{aligned} d_C(P,Q) &= |x_2 - x_1| + |y_2 - y_1| \\ &= |-3-2| + |5-(-3)| \\ &= |-5| + |8| = 13 \text{ blocks} \end{aligned}$$

7. Using the formulas:
$$\begin{aligned} d_E(P,Q) &= \sqrt{(x_2 - x_1)^2 + (y_2 - y_1)^2} \\ &= \sqrt{(5-(-1))^2 + (-2-4)^2} \\ &= \sqrt{36+36} = \sqrt{72} \approx 8.5 \text{ blocks} \end{aligned}$$
$$\begin{aligned} d_C(P,Q) &= |x_2 - x_1| + |y_2 - y_1| \\ &= |5-(-1)| + |-2-4| \\ &= |6| + |-6| = 12 \text{ blocks} \end{aligned}$$

9. Using the formulas:
$$\begin{aligned} d_E(P,Q) &= \sqrt{(x_2 - x_1)^2 + (y_2 - y_1)^2} \\ &= \sqrt{(3-2)^2 + (-6-0)^2} \\ &= \sqrt{1+36} = \sqrt{37} \approx 6.1 \text{ blocks} \end{aligned}$$
$$\begin{aligned} d_C(P,Q) &= |x_2 - x_1| + |y_2 - y_1| \\ &= |3-2| + |-6-0| \\ &= |1| + |-6| = 7 \text{ blocks} \end{aligned}$$

11. Using the formula:
$$\begin{aligned} d_C(P,Q) &= \frac{1+|m|}{\sqrt{1+m^2}}\, d_E(P,Q) \\ &= \frac{1+\left|\frac{3}{4}\right|}{\sqrt{1+\left(\frac{3}{4}\right)^2}} \cdot 5 = \frac{\frac{7}{4}}{\sqrt{\frac{25}{16}}} \cdot 5 \\ &= \frac{\frac{7}{4}}{\frac{5}{4}} \cdot 5 = \frac{7}{4} \cdot \frac{4}{5} \cdot 5 = 7 \text{ blocks} \end{aligned}$$

13. Using the formula:
$$\begin{aligned} d_C(P,Q) &= \frac{1+|m|}{\sqrt{1+m^2}}\, d_E(P,Q) \\ &= \frac{1+\left|-\frac{2}{3}\right|}{\sqrt{1+\left(-\frac{2}{3}\right)^2}} \cdot \sqrt{13} = \frac{\frac{5}{3}}{\sqrt{\frac{13}{9}}} \cdot \sqrt{13} \\ &= \frac{\frac{5}{3}}{\frac{\sqrt{13}}{3}} \cdot \sqrt{13} = \frac{5}{3} \cdot \frac{3}{\sqrt{13}} \cdot \sqrt{13} \\ &= 5 \text{ blocks} \end{aligned}$$

15. Using the formula:
$$\begin{aligned} d_C(P,Q) &= \frac{1+|m|}{\sqrt{1+m^2}}\, d_E(P,Q) \\ &= \frac{1+\left|\frac{1}{4}\right|}{\sqrt{1+\left(\frac{1}{4}\right)^2}} \cdot \sqrt{17} = \frac{\frac{5}{4}}{\sqrt{\frac{17}{16}}} \cdot \sqrt{17} \\ &= \frac{\frac{5}{4}}{\frac{\sqrt{17}}{4}} \cdot \sqrt{17} = \frac{5}{4} \cdot \frac{4}{\sqrt{17}} \cdot \sqrt{17} \\ &= 5 \text{ blocks} \end{aligned}$$

17. A city distance may be associated with more than one Euclidean distance. For example, if $P = (0,0)$ and $Q = (2,0)$, then the city distance between the points is 2 blocks and the Euclidean distance is also 2 blocks. However, if $P = (0,0)$ and $Q = (1,1)$, then the city distance between the points is still 2 blocks, but the Euclidean distance is $\sqrt{2}$ blocks.

19.

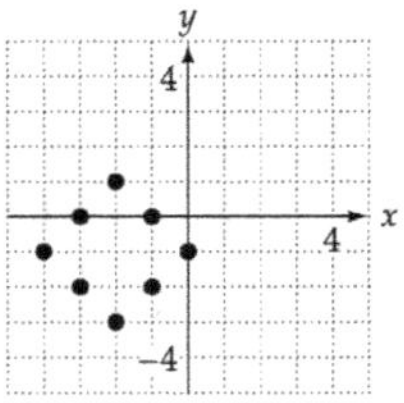

21.

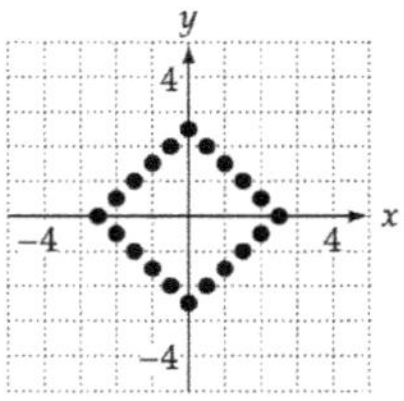

23. $4n$

Think About It, Section 11.8

1. A fractal is a geometric figure in which a self-similar motif repeats itself on an ever-diminishing scale.

2. A fractal is said to be strictly self-similar if any arbitrary portion of the fractal contains a replica of the entire fractal.

Section Exercises 11.8

1.

Stage 2

Stage 3

3.

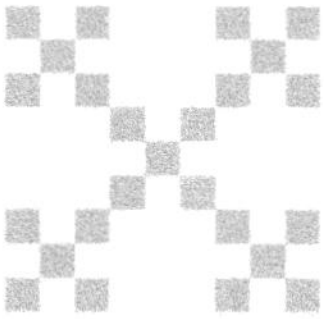

Stage 2

5.

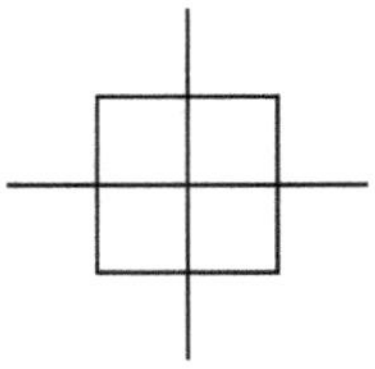

Stage 2

7.

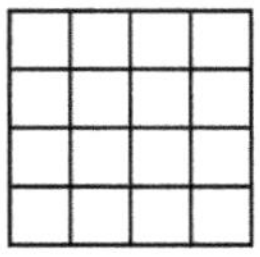

Stage 2

9.

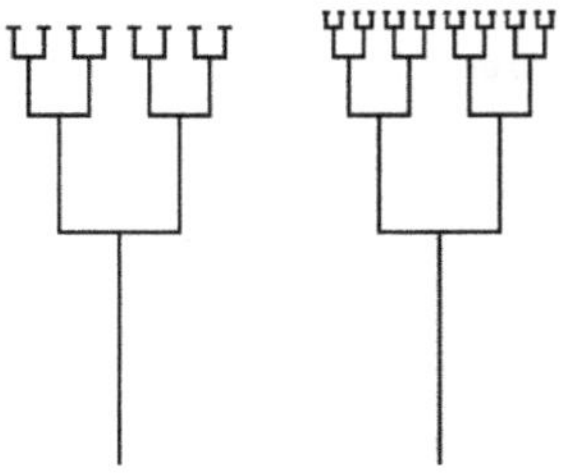

Stage 3 Stage 4

11. Replacement Ratio = 2
    Scale Ratio = 3

    $$\text{Similarity Dimension} = \frac{\log 2}{\log 3} = 0.631$$

13. Replacement Ratio = 5
    Scale Ratio = 3

    $$\text{Similarity Dimension} = \frac{\log 5}{\log 3} = 1.465$$

15. Replacement Ratio = 4
    Scale Ratio = 2

    $$\text{Similarity Dimension} = \frac{\log 4}{\log 2} = 2.000$$

17. Replacement Ratio = 4
    Scale Ratio = 2

    $$\text{Similarity Dimension} = \frac{\log 4}{\log 2} = 2.000$$

19. Replacement Ratio = 18
    Scale Ratio = 6

    $$\text{Similarity Dimension} = \frac{\log 18}{\log 6} = 1.613$$

CHAPTER 11 REVIEW EXERCISES

1. $27 \text{ in.} = 27 \not{\text{in.}} \times \frac{1 \text{ ft}}{12 \not{\text{in.}}} = 2\frac{1}{4} \text{ ft}$

2. $15 \text{ c} = 15 \not{c} \times \frac{1 \text{ pt}}{2 \not{c}} = 7\frac{1}{2} \text{ pt}$

3. 37 mm = 3.7 cm

4. 0.678 g = 678 mg

5. 1273 ml = 1.273 L

6. $\frac{\$3.56}{\text{L}} \approx \frac{\$3.56}{\not{L}} \times \frac{1 \not{L}}{1.06 \text{ qt}} \approx \$3.36/\text{qt}$

7. Calculating:
$m\angle a = 74^\circ \qquad m\angle b = 52^\circ$
$m\angle a = m\angle b + m\angle x$
$74 = 52 + m\angle x$
$22^\circ = m\angle x$

$180^\circ = m\angle x + m\angle y$
$180^\circ = 22 + m\angle y$
$158^\circ = m\angle y$

8. $\frac{AC}{DF} = \frac{BC}{EF}$
$\frac{AC}{12} = \frac{6}{9}$
$72 = 9AC$
Perimeter ABC = 10 + 6 + 8 = 24 in.

9. Break figure into 2 rectangular prisms.
$V_1 = (3)(3)(8) = 72$
$V_2 = (3)(7)(8) = 168$
$V_{\text{total}} = V_1 + V_2$
$V_{\text{total}} = 72 + 168 = 240 \text{ in}^3$

10. $x = 180^\circ - 112^\circ = 68^\circ$

11. Calculating:
$S = 2lw + 2lh + 2wh$
$S = 2(5)(10) + 2(5)(4) + 2(10)(4)$
$S = 220 \text{ ft}^2$

12. $r = \frac{1}{2}(4) = 2 \qquad h = 8$
$S = 2\pi r^2 + 2\pi rh$
$S = 2\pi(2)^2 + 2\pi(2)(8)$
$S = 8\pi + 32\pi = 40\pi \text{ m}^2$

13. $AB = 3BC \qquad BC = 11$
$AC = AB + BC$
$AC = 3(11) + 11 = 44 \text{ cm}$

14. Vertical angles are equal.
$\angle x$ and $\angle y$ are vertical angles.
$m\angle x = 150^\circ$
$m\angle x + m\angle y = 180^\circ$
$150^\circ + m\angle y = 180^\circ$
$m\angle y = 30^\circ$
$m\angle w = 30^\circ$

15. $A = bh$
$A = 6(4.5) = 27 \text{ in}^2$

16. $s = 6 \qquad h = 8$
$V = \frac{1}{3}s^2h$
$V = \frac{1}{3}(6)^2(8) = 96 \text{ cm}^3$

17. $C = 2\pi r = \pi d$
$C = \pi(4.5) \approx 14.1 \text{ m}$

18. $\angle a$ is an alternate interior angle to the given angle. $\angle a$ and $\angle b$ are a linear pair.
$m\angle a = 138^\circ \qquad m\angle b = 42^\circ$

19. $180^\circ - 32^\circ = 148^\circ$
A 148° angle

20. $V = lwh = (6.5)(2)(3) = 39 \text{ ft}^3$

21. Adding:
$m\angle a + m\angle b + m\angle c = 180^\circ$
$37^\circ + 48^\circ + m\angle c = 180^\circ$
$m\angle c = 95^\circ$

22. $A = \frac{1}{2}bh \qquad A = 28 \qquad h = 7$
$28 = \frac{1}{2}b(7) = \frac{7}{2}b$
$b = 8 \text{ cm}$

23. $V = \frac{4}{3}\pi r^3 \qquad r = \frac{1}{2}(12) = 6$
$V = \frac{4}{3}\pi(6)^3 = 288\pi \text{ mm}^3$

24. $P = 4s = 86$
$s = \frac{86}{4} = 21.5 \text{ cm}$

25. $r = 6 h = 15$

$S = 2\pi r^2 + 2\pi rh$

$S = 2\pi(6)^2 + 2\pi(6)(15)$

$S = 252\pi \approx 791.68 \text{ ft}^2$

$791.68 \div 200 = 3.9584$

4 cans of paint are needed

26. $P = 2l + 2w$

$P = 2(56) + 2(48) = 208 \text{ yd}$

27. $A = s^2 = (9.5)^2 = 90.25 \text{ m}^2$

28. $A_{\text{walk}} = A_{\text{total}} - A_{\text{grass}}$

$A_{\text{walk}} = 44(29) + 40(25)$

$A_{\text{walk}} = 1276 - 1000 = 276 \text{ m}^2$

29. Yes, by the SAS Theorem.

30. Using the Pythagorean Theorem:

$a^2 + b^2 = c^2$

$a^2 + 7^2 = 12^2$

$a^2 = 144 - 49 = 95$

$a = \sqrt{95} \approx 9.7 \text{ ft}$

31. Using the Pythagorean Theorem:

$a^2 + b^2 = c^2$

$5^2 + 8^2 = c^2$

$c^2 = 89$

$c = \sqrt{89}$

$\sin\theta = \dfrac{5}{\sqrt{89}} = \dfrac{5\sqrt{89}}{89},$

$\cos\theta = \dfrac{8}{\sqrt{89}} = \dfrac{8\sqrt{89}}{89}, \quad \tan\theta = \dfrac{5}{8}$

32. Using the Pythagorean Theorem:

$a^2 + b^2 = c^2$

$a^2 + 10^2 = 20^2$

$a^2 = 400 - 100 = 300$

$a = \sqrt{300} = 10\sqrt{3}$

$\sin\theta = \dfrac{10\sqrt{3}}{20} = \dfrac{\sqrt{3}}{2},$

$\cos\theta = \dfrac{1}{2}, \quad \tan\theta = \dfrac{10\sqrt{3}}{10} = \sqrt{3}$

33. 25.7°

34. 29.2°

35. 53.8°

36. 1.9°

37. $\tan 50° = \dfrac{d}{84}$

$d = 84 \tan 50° \approx 100.1 \text{ ft}$

38. $\cos 40° = \dfrac{d}{200}$

$d = 200 \cos 40° \approx 153.2 \text{ mi}$

39.

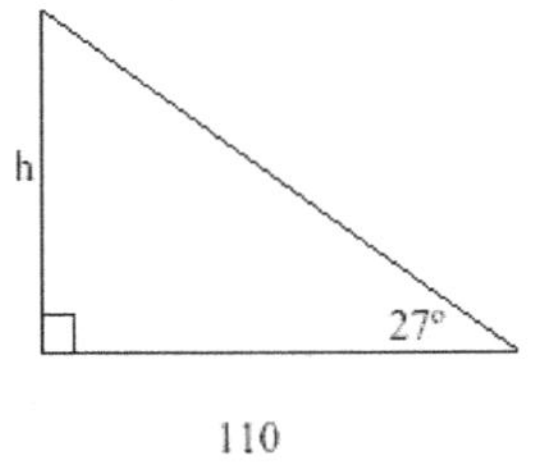

$\tan 27° = \dfrac{h}{110}$

$h = 110 \tan 27° \approx 56.0 \text{ ft}$

40. Spherical geometry or elliptical geometry

41. Hyperbolic geometry

42. Lobachevskian or hyperbolic geometry

43. Riemannian or spherical geometry

44. $m\angle A = 90°, \ m\angle B = 150°,$

$m\angle C = 90°, \ r = 12$

$S = (m\angle A + m\angle B + m\angle C - 180°)\left(\dfrac{\pi}{180°}\right) r^2$

$S = (90° + 150° + 90° - 180°)\left(\dfrac{\pi}{180°}\right)(12)^2$

$S = \dfrac{5}{6}\pi(144) = 120\pi \text{ in}^2$

45. $m\angle A = 90°, \ m\angle B = 60°,$

$m\angle C = 90°, \ r = 5$

$S = (m\angle A + m\angle B + m\angle C - 180°)\left(\dfrac{\pi}{180°}\right) r^2$

$S = (90° + 60° + 90° - 180°)\left(\dfrac{\pi}{180°}\right)(5)^2$

$S = \dfrac{\pi}{3}(25) = \dfrac{25\pi}{3} \text{ ft}^2$

46. Using the formulas:

$d_E(P,Q) = \sqrt{(x_2 - x_1)^2 + (y_2 - y_1)^2}$

$= \sqrt{(3 - (-1))^2 + (4 - 1)^2}$

$= \sqrt{16 + 9} = \sqrt{25} = 5 \text{ blocks}$

$d_C(P,Q) = |x_2 - x_1| + |y_2 - y_1|$

$= |3 - (-1)| + |4 - 1|$

$= |4| + |3| = 7 \text{ blocks}$

47. Using the formulas:

$$d_E(P,Q) = \sqrt{(x_2 - x_1)^2 + (y_2 - y_1)^2}$$
$$= \sqrt{(2-(-5))^2 + (6-(-2))^2}$$
$$= \sqrt{49+64} = \sqrt{113} \approx 10.6 \text{ blocks}$$
$$d_C(P,Q) = |x_2 - x_1| + |y_2 - y_1|$$
$$= |2-(-5)| + |6-(-2)|$$
$$= |7| + |8| = 15 \text{ blocks}$$

48. Using the formulas:

$$d_E(P,Q) = \sqrt{(x_2 - x_1)^2 + (y_2 - y_1)^2}$$
$$= \sqrt{(3-2)^2 + (2-8)^2}$$
$$= \sqrt{1+36} = \sqrt{37} \approx 6.1 \text{ blocks}$$
$$d_C(P,Q) = |x_2 - x_1| + |y_2 - y_1|$$
$$= |3-2| + |2-8|$$
$$= |1| + |-6| = 7 \text{ blocks}$$

49. Using the formulas:

$$d_E(P,Q) = \sqrt{(x_2 - x_1)^2 + (y_2 - y_1)^2}$$
$$= \sqrt{(5-(-3))^2 + (-2-3)^2}$$
$$= \sqrt{64+25} = \sqrt{89} \approx 9.4 \text{ blocks}$$
$$d_C(P,Q) = |x_2 - x_1| + |y_2 - y_1|$$
$$= |5-(-3)| + |-2-3|$$
$$= 8 + |-5| = 13 \text{ blocks}$$

50. a. Using the formulas:

$$d_E(P,Q) = \sqrt{(x_2 - x_1)^2 + (y_2 - y_1)^2}$$
$$= \sqrt{(4-1)^2 + (5-1)^2}$$
$$= \sqrt{9+16} = \sqrt{25} = 5 \text{ blocks}$$
$$d_E(P,R) = \sqrt{(x_2 - x_1)^2 + (y_2 - y_1)^2}$$
$$= \sqrt{(-4-1)^2 + (2-1)^2}$$
$$= \sqrt{25+1} = \sqrt{26} \approx 5.1 \text{ blocks}$$
$$d_E(R,Q) = \sqrt{(x_2 - x_1)^2 + (y_2 - y_1)^2}$$
$$= \sqrt{(4-(-4))^2 + (5-2)^2}$$
$$= \sqrt{64+9} = \sqrt{73} \approx 8.5 \text{ blocks}$$

P and Q are closest.

b. Using the formulas:

$$d_C(P,Q) = |x_2 - x_1| + |y_2 - y_1|$$
$$= |4-1| + |5-1|$$
$$= |3| + |4| = 7 \text{ blocks}$$
$$d_C(P,R) = |x_2 - x_1| + |y_2 - y_1|$$
$$= |-4-1| + |2-1|$$
$$= |-5| + |1| = 6 \text{ blocks}$$
$$d_C(R,Q) = |x_2 - x_1| + |y_2 - y_1|$$
$$= |4-(-4)| + |5-2|$$
$$= |8| + |3| = 11 \text{ blocks}$$

P and R are closest.

51. The Koch curve is a strictly self-similar fractal, because any portion of the Koch curve replicates the entire fractal.

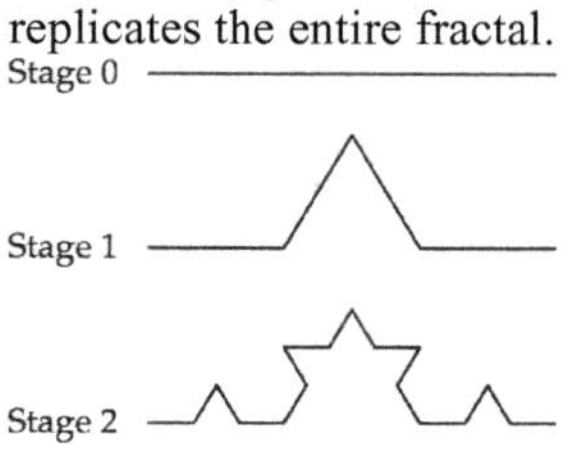

52.

Stage 2

53. a. Replacement Ratio = 2

b. Scaling Ratio = 2

c. Similarity Dimension $= \dfrac{\log 2}{\log 2} = 1$

54. $\dfrac{\log 5}{\log 4} \approx 1.161$

Notetaking Guide & Tips for Success!

We are committed to your success in learning mathematics and are sharing some tips here with you on how to set yourself up for success as well as specifics on how to take notes in class. You may have other tips and tricks you use as well – the key is finding what works best for you!

1. To be successful, attend class! You should consider your commitment to attend class as serious as your commitment to your job or to keeping an appointment with a close friend. It is difficult to overstate the importance of attending class. If you miss work, you don't get paid. If you miss class, you are not getting the full benefit of your tuition dollar. You are losing money.
    - Once in class, participate. Stay involved and active and be present.
    - When your instructor asks a question, try to at least mentally answer the question.
    - If you have a question, ask. Your instructor expects questions and wants you to understand the concept being discussed.
2. If, by some unavoidable situation, you cannot attend class, find out as soon as possible what was covered in class. You might:
    - Ask a friend for notes and the assignment.
    - Contact your instructor and get the assignment. Missing class is no excuse for not being prepared for the next class.
    - Determine whether there are online resources that you can use to help you with the topics and concepts that were discussed in the class you missed.
3. If your instructor uses PowerPoint slides in your classroom:
    - Ask if they are making those available to students after class in an electronic format.
    - Only write down key items or examples during class—don't try and capture everything as that will distract you from learning.
4. Try to reread your notes after class, as regularly as you can, underlining areas you don't understand or need to follow up on with your instructor. The material will be fresh in your mind, and the immediate review, along with your homework, will help reinforce the concepts you are learning.

5. If you can't study or review your notes right after class, make sure that you set aside some time *on the day of the class* to review notes and begin the homework. The longer you wait, the more difficult it will be to recall some of the important points covered during class.
    - Study math in small chunks — *at least* one hour a day, every day, is better than seven hours in one sitting.
    - If you are studying for an extended period of time, break up your study session by studying one subject for a while and then moving on to another subject. Try to alternate between similar or related courses. For instance, study math for a while, then science, and then back to math. Or study history for a while, then political science, and then back to history.

6. Make sure you make time to go to office hours to discuss anything you're unsure about or don't understand.
7. As you prepare for tests, reread your notes and also review your quizzes and homework.

Habits of Successful Students

There are a number of habits that successful students use. Over time, we have asked some of these students to write down the practices they think are important when achieving their course goals. Here are some of their responses:

- **Set priorities.** You will encounter many distractions during the semester. Do not allow them to prevent you from reaching your goal.
- **Take responsibility.** Your instructor, this text, tutors, math centers, and other resources are there to help you succeed. Ultimately, however, you must choose to learn. You must *choose* success.
- **Study with peers.** When you work in a group, you're responsible for each other's success. Seek out people who are committed to success and achieve it together. Success breeds success.
- **Study regularly.** We have mentioned this before, but it is too important not to be repeated.
- **Test yourself.** Once every few days, select homework exercises from previous assignments and use them to test your understanding. Try to do these exercises without getting help from examples in the text. Testing yourself will help you gain confidence that you can do these types of problems on a test given in class.
- **Try different strategies.** If you read the text and are still having difficulty understanding a concept, consider going a step further. Contact the instructor or find a tutor. Many campuses have some free tutorial services. Go to the math or learning center.. Be active and get the help you need.
- **Make flashcards.** This is one of the strategies that some math students do not think to try. Flashcards are a very important part of learning math. For instance, your instructor may use words or phrases such as *linear, quadratic, exponent, base, rational,* and many others. If you don't know the meanings of these words, create flashcards to help you learn them.

Solving Word Problems

One of the reasons you are studying math is to learn to solve word problems. Solving word problems involves combining all of the critical thinking skills you have learned to solve practical problems.

Try not to be intimidated by word problems. Basically, what you need is a strategy that will help you come up with the equation you will need to solve the problem. When you are looking at a word problem, try the following:

- **Read the problem.** This may seem pretty obvious, but we mean really *read* it. Don't just scan it. Read the problem slowly and carefully.
- **Write down what is known and what is unknown.** Now that you have read the problem, go back and write down everything that is known. Next, write down what it is you are trying to find. *Write* this — don't just think it! Be as specific as you can.

For instance, if you are asked to find a distance, instead of writing "I need to find the distance," be more specific and write "I need to find the distance *between Earth and the moon*."

- **Think of a method to find the unknown.** For instance, is there a formula that relates the known and unknown quantities? Identifying one will help you write an equation to be solved.
- **Solve the equation.** Be careful as you solve the equation.. The unknown in most word problems will include a unit such as feet, dollars, or miles per hour. When you write your answer, include the unit. An answer such as 20 doesn't mean much. Is it 20 feet, 20 dollars, 20 miles per hour, or something else?
- **Check your solution.** Now that you have an answer, go back to the problem and ask yourself whether it is reasonable and makes sense. This is an important step. For instance, if, according to your answer, the cost of a car is $2.51, you know that something went wrong!

Ready, Set, Succeed!

It takes hard work and commitment to succeed, but we know you can do it! Doing well in mathematics is just one step you'll take on your path to success. Good luck. We wish you success.

Notetaking Template

The template below is a recommendation of how to organize your notes. Please view it as a starting point that you can tailor to your needs.

Topic	Chapter/Section	Key points	Formulas/Examples/Problems
Example: Credit Cards	Ch4, Section 4.4	Paying off a credit card	Formula: Present Value of an Ordinary Annuity $PV = PMT\left[\frac{1-(1+i)^{-n}}{i}\right]$

CPSIA information can be obtained
at www.ICGtesting.com
Printed in the USA
BVHW010255111022
649121BV00019B/156

9 780357 028063